HUTCHINSON POCKET

Dictionary of Biology

Helicon

Helicon Publishing Ltd
42 Hythe Bridge Street
Oxford OX1 2EP

Typeset by Roger Walker Graphic Design and Typography,
Maidenhead, Berkshire

Printed and bound in Great Britain by
Unwin Brothers Ltd, Old Woking, Surrey

ISBN 0 09 178102 7

British Cataloguing in Publication Data

A catalogue record for this book is available
from the British Library

Editorial director
Michael Upshall

Consultant editor
Stephen Webster BSc, MPhil

Project editor
Sara Jenkins-Jones

Text editor
Catherine Thompson

Art editor
Terence Caven

Additional page make-up
Helen Bird

Production
Tony Ballsdon

A

abdomen the part of the body below the ◊thorax, containing the stomach, intestines, liver, and kidneys; in insects and other arthropods, it is the hind part of the body. In mammals, the abdomen is separated from the thorax by the ◊diaphragm, a sheet of muscular tissue.

abiotic factor a non-living variable within the ◊ecosystem that affects the life of organisms. Examples include temperature, light, and ◊soil structure and composition. Abiotic factors can be harmful to the environment, as when sulphur dioxide emissions from power stations produce acid rain.

abscission the controlled separation of a part of a plant from the main plant body—most commonly, the falling of leaves or the dropping of fruit. In ◊deciduous plants the leaves are shed before the winter or dry season, whereas ◊evergreen plants drop their leaves continually throughout the year. The process is thought to be controlled by the amount of ◊auxin (growth hormone) present.

absorption the uptake of materials by organisms for use in their cellular processes. Any molecule small enough to pass through the pores of a semipermeable membrane can be absorbed by a cell, although the rate and extent of movement will depend on the relative concentrations of the molecules inside and outside the cell. Generally, molecules are absorbed by ◊diffusion from a region of high concentration to a region of low concentration; in some instances, however, molecules are transported actively from a region of low concentration to one of high concentration (see ◊active transport).

A simple example of absorption is found in the human lungs, where oxygen passes from the air sacs, or alveoli, into the blood capillaries by diffusing through a thin membrane. Absorption also occurs in the gut (alimentary canal) where small molecules of nutrient, such as sugars or amino acids, diffuse (or are actively transported) across the gut wall into the blood system.

accommodation the ability of the vertebrate ⟐eye to focus on near or far objects by changing the shape of the lens.

For something to be viewed clearly the image must be precisely focused on the retina, the light-sensitive sheet of cells at the rear of the eye. Close objects can be seen when the lens takes up a more spherical shape, far objects when the lens is stretched and made thinner. These changes in shape are directed by the brain and by a ring of ciliary muscles lying beneath the iris.

acid rain acidic rainfall, thought to be caused principally by the release into the atmosphere of sulphur dioxide (SO_2) and oxides of nitrogen. Sulphur dioxide is formed from the burning of fossil fuels, such as coal, that contain high quantities of sulphur; nitrogen oxides are contributed from various industrial activities and from car exhaust fumes.

Acid rain is linked with damage to and death of forests and lake organisms in Scandinavia, Europe, and eastern North America. It also results in damage to buildings and statues.

acquired character a feature of the body that develops during the lifetime of an individual, usually as a result of repeated use or disuse, such as the enlarged muscles of a weightlifter. Lamarck's theory of evolution (see ⟐Lamarckism) assumed that acquired characters were passed from parent to offspring.

Modern evolutionary theory does not recognize the inheritance of acquired characters because there is no reliable scientific evidence that it occurs, and because no mechanism is known whereby bodily changes can influence the genetic material.

active transport the use of energy to move molecules or ions across a cell membrane and against a concentration gradient (from a region of low concentration to a region of high concentration). The process is thought to involve the binding of the molecule to be absorbed with a protein carrier molecule in the cell membrane. Examples of active transport include, in animals, the absorption of amino acids by cells in the ileum wall and, in plants, the taking up of minerals by root hairs.

Active transport may be contrasted with ◊diffusion, a passive process (requiring no input of energy) by which molecules are absorbed into a region in which they are in lower concentration, as when oxygen passes from the alveoli of the lungs into the blood capillaries.

adaptation any feature of an organism that allows it to survive and reproduce more effectively in its environment. The theory of evolution by natural selection holds that species become extinct when they are no longer adapted to their environment—for instance, if the climate becomes suddenly colder. By definition, all organisms existing now must be reasonably well adapted.

adaptive radiation in evolution, the formation of several species, with adaptations to different ways of life, from a single ancestral type. Adaptive radiation is likely to occur whenever members of a species migrate to a new habitat, such as an island, with unoccupied ecological ◊niches. It is thought that the lack of competition in such niches allows sections of the migrant population to develop new adaptations, and eventually to become new species.

additive in food technology, any natural or artificial chemical added to prolong the shelf life of processed foods (salt or nitrates), alter the colour or flavour of food, or improve its food value (vitamins or minerals). Many chemical additives are used and they are subject to regulation, since individuals may be affected by constant exposure even to traces of certain additives and may suffer side effects ranging from headaches and hyperactivity to cancer. Within the European Community, approved additives are given an official ***E number***.

flavours are said to increase the appeal of the food. They may be natural or artificial, and include artificial sweeteners and monosodium glutamate (m.s.g.).

colourings are used to enhance the visual appeal of certain foods.

enhancers are used to increase or reduce the taste and smell of a food without imparting a flavour of their own.

nutrients replace or enhance food value. Minerals and vitamins are added if the diet might otherwise be deficient, to prevent diseases such as beri-beri and pellagra.

adaptation

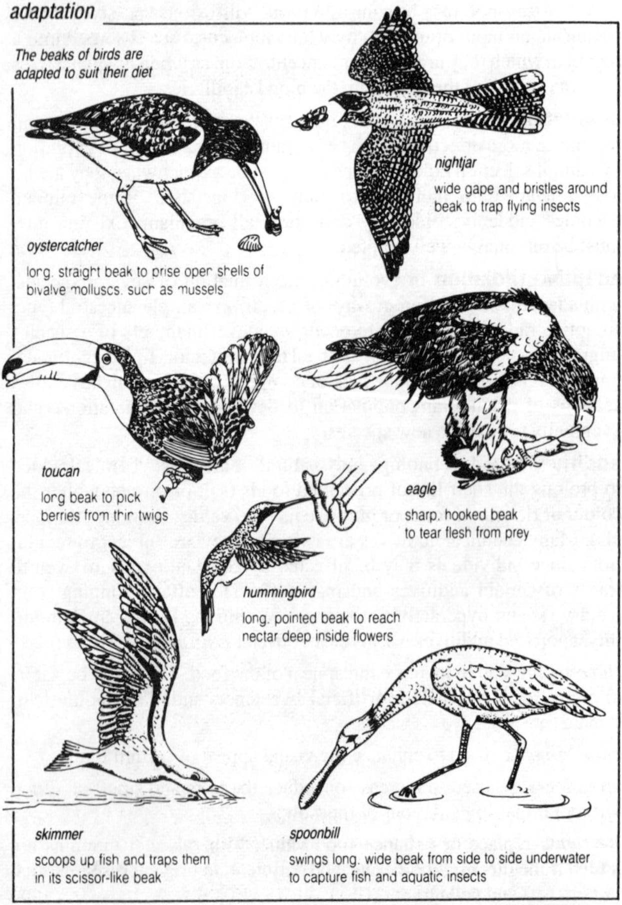
The beaks of birds are adapted to suit their diet
oystercatcher
long, straight beak to prise open shells of bivalve molluscs, such as mussels
nightjar
wide gape and bristles around beak to trap flying insects
toucan
long beak to pick berries from thin twigs
eagle
sharp, hooked beak to tear flesh from prey
hummingbird
long, pointed beak to reach nectar deep inside flowers
skimmer
scoops up fish and traps them in its scissor-like beak
spoonbill
swings long, wide beak from side to side underwater to capture fish and aquatic insects

preservatives are antioxidants and antimicrobials that control natural oxidation and the action of microorganisms. See ◊food technology.

emulsifiers and ***surfactants*** regulate the consistency of fats in the food and on the surface of the food in contact with the air.

thickeners, primarily vegetable gums, regulate the consistency of food. Pectin acts in this way on fruit products.

leavening agents lighten the texture of baked goods without the use of yeasts. Sodium bicarbonate is an example.

acidulants sharpen the taste of foods but may also perform a buffering function in the control of acidity.

bleaching agents assist in the ageing of flours.

anti-caking agents prevent powdered products coagulating into solid lumps.

humectants control the humidity of the product by absorbing and retaining moisture.

clarifying agents are used in fruit juices, vinegars, and other fermented liquids. Gelatin is the most common.

firming agents restore the texture of vegetables that may be damaged during processing.

foam regulators are used in beer to provide a controlled 'head' on top of the poured product.

ADH (abbreviation for ***antidiuretic hormone***) hormone secreted by the ◊pituitary gland that plays a role in maintaining a correct salt/water balance in the blood. It stimulates the kidneys to conserve water more efficiently, thereby allowing the body to compensate for a varying water intake.

In conditions of water shortage, the concentration of the blood is raised triggering receptors, which in turn bring about the secretion of ADH by the pituitary. The resulting high levels of ADH in the blood stimulate the kidneys to produce more concentrated urine so that less water is lost from the body. When the animal is able to drink plenty of water, decreased ADH secretion causes the kidneys to produce more dilute urine so that more water leaves the body and the blood does not become too dilute.

adipose tissue a type of ◊connective tissue that serves as an energy reserve, and also pads some organs. It is commonly called fat tissue, and consists of large spherical cells filled with fat. In mammals, major layers are in the inner layer of skin and around the kidneys and heart.

adolescence in the human lifecycle, the period between the beginning of puberty and adulthood.

ADP abbreviation for ***adenosine diphosphate***, a raw material in the manufacture of ◊ATP (adenosine triphosphate), the molecule used by all cells to drive their metabolic reactions.

adrenal gland or ***suprarenal gland*** a gland situated on top of the kidney. The adrenals are soft and yellow, and consist of two parts: the outer cortex and the inner medulla. The ***cortex*** secretes various hormones, controls salt and water metabolism, and regulates the use of carbohydrates, proteins, and fats. The ***medulla*** secretes the hormones adrenaline and noradrenaline, which constrict the blood vessels of the belly and skin so that more blood is available for the heart, lungs, and voluntary muscles in emergency 'fight or flight' situations.

adrenaline or ***epinephrine*** hormone secreted by the medulla of the ◊adrenal glands.

aerobe organism that respires aerobically. Almost all living organisms (plants as well as animals) are aerobes, and will die in the absence of oxygen. Compare ◊anaerobe.

aerobic respiration form of respiration that requires the presence of oxygen (usually dissolved in water) for the efficient release of energy contained in food molecules, such as glucose.

Aerobic respiration occurs inside the ◊mitochondria of the cell and, unlike ◊anaerobic respiration, involves the complete breakdown of glucose to give carbon dioxide, water, and large amounts of energy (stored in the form of ◊ATP molecules), which will subsequently be used by the cell for driving its metabolic processes.

$$C_6H_{12}O_6 + O_2 \rightarrow 6CO_2 + 6H_2O + 2{,}880\ \text{kJ}$$

afterbirth the placenta and other material, including blood and membranes, expelled from the mammalian uterus soon after birth. In the natural world it is often eaten.

agar jellylike substance, obtained from seaweeds. It is used mainly in microbiological experiments as a culture medium for growing bacteria and other microorganisms.

agglutination the clumping together of ◊antigens, such as blood cells or bacteria, to form larger, visible masses, under the influence of ◊antibodies. As each antigen clumps only in response to its particular antibody, agglutination provides a way of determining ◊blood groups and the identity of unknown bacteria. See ◊immunity.

aggression behaviour used to intimidate or injure another organism (of the same or of a different species), usually for the purposes of gaining a territory, a mate, or food. Aggression often involves an escalating series of threats aimed at intimidating an opponent without having to engage in potentially dangerous physical contact ('fights to the death' are rare in nature). Aggressive signals include roaring by red deer, snarling by dogs, the fluffing up of feathers by birds, and the raising of fins by some species of fish.

agriculture the cultivation of land and the raising of domesticated animals in order to provide food or materials such as wool and cotton. The development of agriculture was a significant step in the history of humankind. Previously, food had been obtained only by hunting and by gathering wild vegetation. The selective breeding (see ◊artificial selection) of reliable and productive animals and crop plants and the improvement of soil by ploughing, irrigation, ◊crop rotation, and the use of organic ◊fertilizers, such as manure and ashes, meant that communities could become more stable, giving rise to fixed villages and towns and to complicated social systems. The increased demands made upon agriculture by the growing world population have led to greater land clearance and the intensification of farming methods, which now include the use of chemical pesticides such as ◊herbicides (weedkillers) and ◊insecticides and of artificial, non-organic fertilizers.

In the 20th century some of the methods used in agriculture have caused concern. Land clearance and ◊deforestation has destroyed the natural habitats of many animal and plant species (see ◊endangered species and ◊extinction) and has also led to ◊soil erosion in which the top, fertile layer of soil—no longer anchored by tree and shrub roots—is blown or washed away, leaving behind a barren desert or 'dust bowl'. Fertilizers can leach away from the soil to pollute water supplies and aquatic ecosystems, and pesticides can pass through food chains, accumulating in the diets of animals, including humans, at the highest trophic levels. The intensive rearing (factory farming) of animals such as pigs and chickens has lowered the cost of meat, but has also aroused controversy about its cruelty and about possible health hazards such as salmonella food poisoning.

AIDS (acronym for ***acquired immune deficiency syndrome***) the newest and gravest of the sexually transmitted diseases (STDs). It is caused by the ***human immunodeficiency virus*** (HIV), which is transmitted in body fluids—mainly blood and sexual secretions. Sexual transmission of the AIDS virus endangers heterosexual men and women as well as high-risk groups, such as homosexual and bisexual men, prostitutes, intravenous drug-users sharing needles, and haemophiliacs and surgical patients treated with contaminated blood products. The virus has a short life outside the body, which makes its transmission by methods other than sexual contact, blood transfusion, and shared syringes extremely unlikely.

Infection with HIV does not necessarily mean that a person has AIDS: many people who have the virus in their blood are not ill; others suffer AIDS-related illnesses but not the full-blown disease. The effect of the virus in those who become ill is the devastation of the immune system, leaving the victim susceptible to diseases that would not otherwise develop. Some AIDS sufferers die within a few months of the outbreak of symptoms, some survive for several years; roughly 50% are dead within three years. There is no cure for the disease, and the search continues for an effective vaccine.

Global Aids Policy Coalition, Harvard University, USA, published a report *Aids in the World* 1993 in which it stated that 12.9 million people worldwide had been infected with HIV by the end of 1992. The same report predicts that 20 million people will be infected by 1995 and that 38–110 million adults and over 10 million children will be infected by the year 2000, with 25 million full-blown AIDS cases worldwide by the year 2000.

air pollution contamination of the atmosphere caused by the discharge, accidental or deliberate, of a wide range of toxic substances such as sulphur dioxide and carbon monoxide. Often the amount of the released substance is relatively high in a certain locality, so the harmful effects are more noticeable. The cost of preventing any discharge of pollutants into the air is very high, so attempts are more usually made to reduce gradually the amount of discharge and to disperse this as quickly as possible by using a very tall chimney, or by intermittent release.

alimentary canal another word for ◊gut, the tube through which food passes, and in which it is processed, digested, and absorbed.

allele one of two or more alternative forms of a ◊gene at a given locus on a chromosome, caused by a difference in the ◊DNA. Blue and brown eyes in humans are determined by different alleles of the gene for eye colour.

Organisms with two sets of chromosomes (diploids) will have two copies of each gene. If the two alleles are identical the individual is said to be ◊homozygous at that ◊locus; if different, the individual is ◊heterozygous at that locus. Some alleles show dominance over others; see ◊dominant and ◊recessive.

allometry regular relationship between a given feature (for example, the size of an organ) and the size of the body as a whole, when this relationship is not a simple proportion of body size. Thus, an organ may increase in size proportionately faster, or slower, than body size does.

The best known allometric relationship is the ***surface-area law***: the ratio of body surface to total body volume decreases as body size gets larger. Large animals therefore lose less heat than small ones because they have proportionately less skin surface from which to radiate heat.

alternation of generations the life cycle of land plants and some seaweeds in which there are two distinct forms occurring alternately:

diploid (having two sets of chromosomes) and ***haploid*** (one set of chromosomes). The diploid generation produces haploid spores by ◊meiosis, and is called the sporophyte, while the haploid generation produces gametes (sex cells), and is called the gametophyte. The gametes fuse to form a diploid ◊zygote which develops into a new sporophyte; thus the sporophyte and gametophyte alternate.

The life cycles of certain animals (such as the jellyfish) are sometimes said to show alternation of generations, but this is rarely as regular and clearly-defined as in plants.

alternative energy energy from sources that are renewable and ecologically safe, as opposed to sources that are nonrenewable with toxic by-products, such as coal, oil, or gas (fossil fuels), and uranium (for nuclear power). The most important alternative energy source is flowing water, harnessed as hydroelectric power. Other sources include the ocean's tides and waves, the wind (harnessed by windmills and wind turbines), the Sun (solar energy), and the heat trapped in the Earth's crust (geothermal energy).

alveolus one of the many thousands of tiny air sacs in the ◊lungs in which exchange of oxygen and carbon dioxide takes place between air and the bloodstream. See ◊respiratory surface and ◊gas exchange.

alveolus

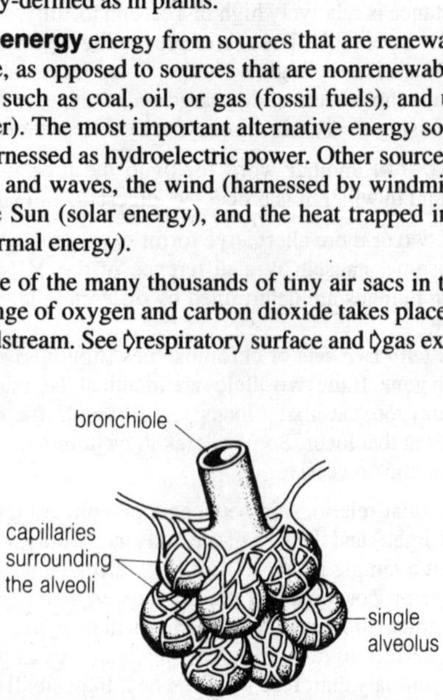

amino acid water-soluble organic molecule, mainly composed of carbon, oxygen, hydrogen, and nitrogen, containing both a basic amine group ($-NH_2$) and an acidic carboxyl ($-COOH$) group. When two or more amino acids are joined together, they are known as ◊peptides; ◊proteins are made up of interacting polypeptides (peptide chains consisting of more than three amino acids) and are folded or twisted in characteristic shapes.

Many different proteins are found in the cells of living organisms, but they are all made up of the same 20 amino acids, joined together in varying combinations, (although other types of amino acid do occur infrequently in nature). Eight of these, the ***essential amino acids***, can not be synthesized by humans and must be obtained from the diet. Children need a further two amino acids that are not essential for adults. Other animals also need some preformed amino acids in their diet, but green plants can manufacture all the amino acids they need from simpler molecules, relying on energy from the Sun and minerals (including nitrates) from the soil.

ammonification process that takes place in the soil in which bacterial and fungal decomposers (saprotrophs) convert proteins from dead organisms and urea from excrement into ammonia. See ◊nitrogen cycle.

amnion in mammals, the innermost of the three membranes that enclose the fetus within the uterus. It contains the ***amniotic fluid***, which helps to cushion the embryo.

amoeba (plural ***amoebae***) one of the simplest living animals, consisting of a single cell and belonging to the ◊protozoa. The body consists of colourless protoplasm. Its activities are controlled by its nucleus, and it feeds by flowing round and engulfing organic debris. It reproduces by ◊binary fission. Some of its relatives are harmful parasites.

amphibian (Greek 'double life') member of a class of vertebrates (Amphibia), which generally spend their larval (tadpole) stage in fresh water and undergo ◊metamorphosis, transferring to land at maturity and usually returning to water to breed. Like fish and reptiles, they continue to grow throughout life and cannot maintain a temperature greatly differing from that of their environment. The class includes salamanders, newts, frogs, and toads.

ampulla in the ◊ear, the slight swelling at the end of each semicircular canal, able to sense the motion of the head. The sense of balance largely depends on sensitive hairs within the ampulla responding to movements of fluid within the inner ear.

amylase one of a group of ◊enzymes that breaks down ◊starches into their component molecules (sugars) for use in the body. It occurs widely in both plants and animals. In humans, it is found in saliva and in pancreatic juices.

anabolism the process of building up body tissue, promoted by the influence of certain hormones. It is the constructive side of ◊metabolism, as opposed to catabolism.

anaemia condition caused by a shortage of haemoglobin, the oxygen-carrying component of red blood cells. The main symptoms are fatigue, paleness of the skin, breathlessness, palpitations, and poor resistance to infection.

Anaemia arises either from abnormal loss or poor production of haemoglobin. Excessive loss occurs, for instance, with chronic slow bleeding or with accelerated destruction, or ◊haemolysis, of red blood cells. Causes of defective production include iron or cyanocobalamine (vitamin B_{12}) deficiency, malnutrition, and blood diseases such as ◊sickle-cell disease. Untreated anaemia puts a strain on the heart and may prove fatal.

anaerobe organism that respires anaerobically. Anaerobes include many bacteria, yeasts, and internal parasites.

Obligate anaerobes, such as certain primitive bacteria, cannot function in the presence of oxygen; but ***facultative anaerobes***, like the fermenting yeasts and most bacteria, respire aerobically in the presence of oxygen, and anaerobically in its absence. Compare ◊aerobe.

anaerobic respiration form of respiration that does not require the presence of oxygen for the release of energy from food molecules, such as glucose. Because the food molecule is only partially broken down in the process, anaerobic respiration releases 19 times less of the molecule's available energy than does ◊aerobic respiration.

In yeasts and bacteria, anaerobic respiration involves the breakdown of glucose to give ethanol (alcohol), carbon dioxide, and energy (stored in the form of ◊ATP molecules).

$$C_6H_{12}O_6 \rightarrow 2C_2H_5OH + 2CO_2 + 210\text{ kJ}$$

The process is exploited by both the brewing and the baking industries (see ◊fermentation).

Normally aerobic animal cells can breathe anaerobically for short periods of time when oxygen levels are low, breaking down glucose to give lactic acid and energy.

$$C_6H_{12}O_6 \rightarrow 2C_3H_6O_3 + 150\text{ kJ}$$

However, the cells involved are ultimately fatigued by the build-up of lactic acid. This form of respiration is seen particularly in muscle cells during intense activity, when the demand for oxygen can outstrip supply (see ◊oxygen debt).

analgesic drug that relieves ◊pain. It differs from an anaesthetic in that it does not cause loss of sensation or of consciousness. Analgesics include opiates, such as morphine, which are effective in controlling 'deep' internal pain, and non-opiates, such as aspirin and paracetamol, which relieve muscular pain and reduce inflammation in soft tissues.

analogous term describing a structure or organ that is similar in function to a structure or organ in another organism, but not a similar evolutionary path. For example, the wings of bees and of birds have the same purpose—to give powered flight—but have different origins.

In contrast, ◊homologous structures such as the dolphin flipper and the digging forelimbs of a mole have different functions but are clearly evolutionarily related; they share a similar position in the vertebrate tetrapod format.

anatomy the study of the structure of the body and its component parts, especially the ◊human body, as distinguished from physiology, which is the study of bodily functions.

androecium the male part of a flower, comprising a number of ◊stamens.

androgen general name for any male sex hormone, of which ◊testosterone is the most important. They are principally involved in the production of male ◊secondary sexual characters (such as facial hair in humans).

anemophily a type of ◊pollination in which the pollen is carried on the wind.

angiosperm flowering plant in which the seeds are enclosed within an ovary, which ripens to a fruit. Angiosperms are divided into ◊monocotyledons (single seed leaf in the embryo) and ◊dicotyledons (two seed leaves in the embryo). They include the majority of flowers, herbs, grasses, and trees (except conifers).

animal member of the kingdom Animalia, one of the major categories of living things, the science of which is ***zoology***. Animals are all ◊heterotrophs (they obtain their energy from organic substances produced by other organisms); they have ◊eukaryotic cells (the genetic material is contained within a distinct nucleus) bounded by a thin cell membrane rather than the thick cell wall of plants. In the past, it was common to include the single-celled ◊protozoa with the animals, but these are now classified as protists, together with single-celled plants. Thus all animals are multicellular. Most are capable of moving around for at least part of their life cyle.

annual plant a plant that completes its life cycle within one year, during which time it germinates, grows to maturity, bears flowers, produces seed, and then dies. Examples include the common poppy and groundsel. See also ◊biennial plant and ◊perennial plant.

annual rings or ***growth rings*** concentric rings visible on the wood of a cut tree trunk or other woody stem. Each ring represents a period of growth when new ◊xylem is laid down to replace tissue being converted into wood (secondary xylem). The wood formed from xylem produced in the spring and early summer has larger and more numerous vessels than the wood formed from xylem produced in autumn when growth is slowing down. The result is a clear boundary between the pale spring wood and the denser, darker autumn wood. Annual rings may be used to estimate the age of the plant, although occasionally more than one growth ring is produced in a given year.

annual rings

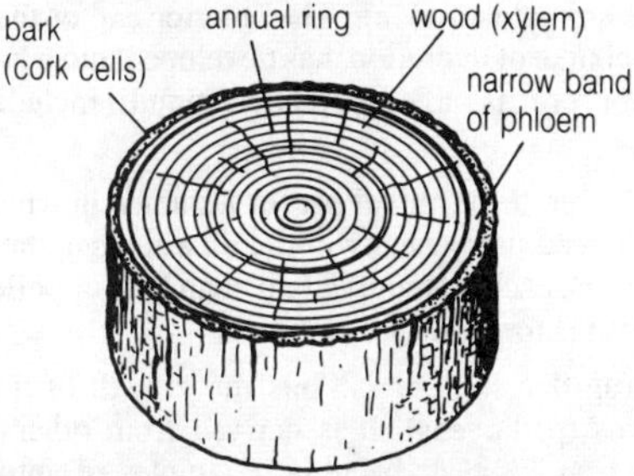

antagonistic muscle pair of muscles allowing cordinated movement of the skeletal joints. The extension of the arm, for example, requires one set of muscles to relax, while another set contracts. The individual components of antagonistic pairs can be classified into ◊extensors and ◊flexors.

antagonistic muscle

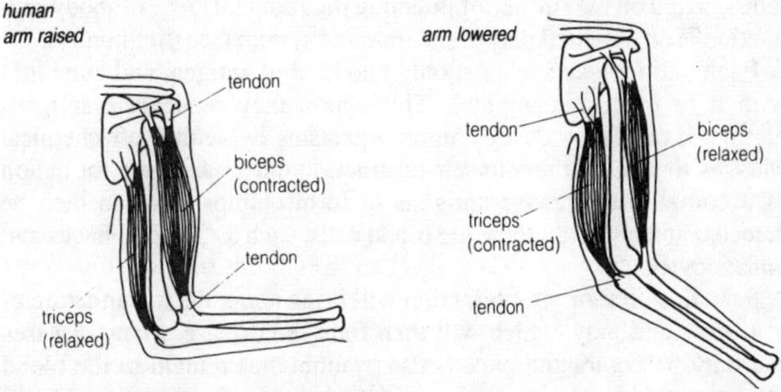

antenna in zoology, an appendage ('feeler') on the head. In insects the antennae are usually involved with the senses of smell and touch.

anterior the front of an organism, usually the part that goes forward first when the animal is moving. The anterior end of the nervous system, over the course of evolution, has developed into a brain with associated receptor organs able to detect stimuli including light and chemicals.

anther in a flower, the terminal part of a stamen in which the ◊pollen grains are produced. It is usually borne on a slender stalk or filament, and has two lobes, each containing two chambers or pollen sacs within which the pollen is formed.

antibiotic drug that kills or inhibits the growth of disease-causing microorganisms (pathogens). It is derived from other living microorganisms such as fungi or bacteria. Examples of antibiotic include penicillin (the first to be discovered) and streptomycin.

Each type of antibiotic acts in a different way and may be effective against either a broad range or a specific type of pathogen. New mutant strains of bacteria have the ability to develop ***resistance*** to antibiotics, so more advanced antibiotics are continually required in order to overcome them.

antibody proteins produced in the body in response to the presence of foreign or invading substances, or ◊antigens, which include the proteins carried on the surface of infecting microorganisms. Antibody production is an essential part of the immune system (see ◊immunity).

Each antibody acts against only one kind of antigen, and combines with it to form a 'complex'. This action may render the antigens harmless, or it may destroy microorganisms by setting off chemical changes that cause them to self-destruct. In other cases, the formation of a complex will cause antigens to form clumps that can then be detected and engulfed by white blood cells, such as ◊macrophages and ◊phagocytes.

Each bacterial or viral infection will bring about the manufacture of a specific antibody, which will then fight the disease. Many diseases can only be contracted once because antibodies remain in the blood after the infection has passed, preventing any further invasion. Vaccination boosts a person's resistance by causing the production of antibodies specific to particular infections.

antigen any substance that causes production of ◊antibodies. Common antigens include the proteins carried on the surface of bacteria, viruses, and pollen grains. The proteins of incompatible blood groups or tissues also act as antigens, which has to be taken into account in medical procedures such as blood transfusions and organ transplants.

antioxidant type of food ◊additive, used to prevent fats and oils from becoming rancid, and thus extend their shelf life.

antiseptic any chemical that kills or inhibits the growth of microorganisms.

anus opening at the end of the ◊gut that allows undigested food and associated materials to pass out of the animal.

aorta the chief ◊artery, the dorsal blood vessel carrying oxygenated blood from the left ventricle of the heart. It branches to form smaller arteries and arterioles, which in turn supply all body organs except the lungs.

appendix area of the mammalian gut, associated with the digestion of ◊cellulose. In herbivores it may be large, containing millions of bacteria secreting enzymes to digest plant material. The human appendix is a ◊vestigial organ, serving no digestive function.

aquatic living in water. All life originated in the early oceans because the aquatic environment has several advantages for organisms: dehydration is almost impossible, temperatures usually remain stable, and the heavyness of water provides physical support.

aqueous humour watery fluid found in the space between the cornea and lens of the ◊eye.

arteriosclerosis hardening and thickening of the arteries due to the deposition of substances such as fatty acids, cholestrol, and calcium. The condition lowers the elasticity of the artery walls, and can lead to high blood pressure, loss of circulation, heart disease, and death. It is associated with smoking, ageing, and a diet high in saturated fats (see ◊polyunsaturate).

artery vessel that carries blood from the heart to the rest of the body. It is built to withstand considerable pressure, having thick walls that are impregnated with muscle and elastic fibres. During contraction of the heart muscles, arteries expand in diameter to allow for the sudden increase in pressure that occurs; the resulting ◊pulse or pressure wave can be felt at the wrist. Not all arteries carry oxygenated (oxygen-rich) blood; the pulmonary arteries convey deoxygenated (oxygen-poor) blood from the heart to the lungs.

Arteries are vulnerable to damage by the build-up of deposits of substances such as fatty acids and cholestrol—a condition known as ◊arteriosclerosis.

articulating surface surface of a bone that rubs against or glides over the surface of another in a slightly movable ◊joint. Such surfaces are covered with a protective layer of ◊cartilage.

artificial selection selective breeding of individuals that exhibit the particular characteristics that a plant or animal breeder wishes to develop. In plants, desirable features might include resistance to disease, high yield (in crop plants), or attractive appearance. In animal breeding, selection has led to the development of particular breeds of cattle for improved meat production (such as the Aberdeen Angus) or milk production (such as Jerseys).

ascorbic acid or ***vitamin C*** relatively simple organic acid found in fresh fruits and vegetables. It is soluble in water and destroyed by prolonged boiling, so soaking or overcooking of vegetables reduces their vitamin C content.

In the human body, ascorbic acid is necessary for the correct synthesis of ◊collagen. Lack of it causes skin sores or ulcers, tooth and gum problems, and burst capillaries (all symptoms of the deficiency disease scurvy). See ◊vitamin.

asexual reproduction reproduction that does not involve the manufacture and fusion of sex cells, nor the necessity for two parents. The process carries a clear advantage in that there is no need to search for a mate nor to develop complex pollinating mechanisms; every asexual organism can reproduce on its own. Asexual reproduction can therefore lead to a rapid population build-up.

In evolutionary terms, the disadvantage of asexual reproduction arises from the fact that only identical individuals, or clones, are produced—

asexual reproduction

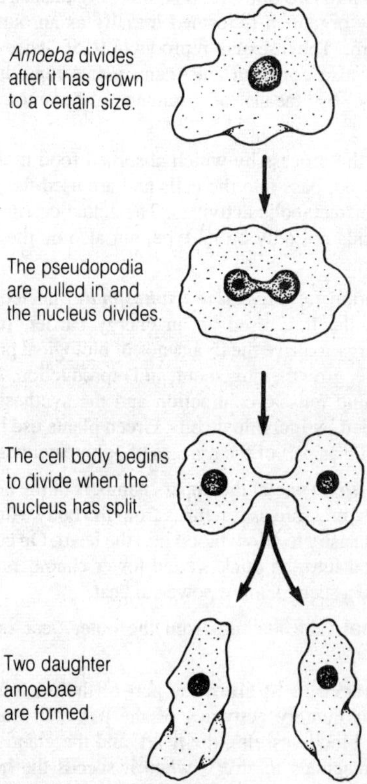

there is no variation. In the field of horticulture, where standardized production is needed, this is useful, but in the wild, an asexual population that cannot adapt to a changing environment is at risk of extinction.

Asexual processes include ◊binary fission, in which the parent organism splits into two or more 'daughter' organisms, and ◊budding, in which a new organism is formed initially as an outgrowth of the parent organism. The asexual reproduction of spores, as in ferns and mosses, is also common and many plants reproduce asexually, or vegetatively, by means of ◊runners, ◊rhizomes, ◊bulbs, and ◊corms.

assimilation the process by which absorbed food molecules, circulating in the blood, pass into the cells and are used for growth, tissue repair, and other metabolic activities. The actual destiny of each food molecule depends not only on its type, but also on the body requirements at that time.

ATP (abbreviation for ***adenosine triphosphate***) nucleotide molecule found in all cells. It is used as an energy carrier, releasing large amounts of energy to drive the thousands of biological processes needed to sustain life, growth, movement, and reproduction. ATP is the driving force behind muscle contraction and the synthesis of complex molecules needed by individual cells. Green plants use light energy to manufacture ATP as part of the process of ◊photosynthesis.

atrium (plural ***atria***) one of the upper chambers of the heart, receiving blood under low pressure as it returns from the body. Atrium walls are thin and stretch easily to allow blood into the heart. On contraction, the atria force blood into the thick-walled lower chambers (◊ventricles), which then give a second, more powerful beat.

auditory canal tube leading from the outer ◊ear opening to the eardrum.

autonomic nervous system the part of the nervous system that controls the involuntary activities of the ◊smooth muscles (of the digestive tract, blood vessels), the heart, and the glands. The ***sympathetic*** system responds to stress, when it speeds the heart rate, increases blood pressure and generally prepares the body for action. The ***parasympathetic*** system is more important when the body is at rest, since it slows the heart rate, decreases blood pressure, and stimulates the digestive system.

autoradiography technique for following the movement of molecules within an organism, especially a plant, by labelling them with a radioactive isotope that can be traced on photographs. It is used to study ◊photosynthesis, where the pathway of radioactive carbon dioxide can be traced as it moves through the various chemical stages.

autosome any ◊chromosome in the cell other than a sex chromosome. Autosomes are of the same number and kind in both males and females of a given species.

autotroph any living organism that synthesizes organic substances from inorganic molecules by using light or chemical energy. Autotrophs are the ***primary producers*** in all food chains since the materials they synthesize and store are the energy sources of all other organisms. All green plants and many planktonic organisms are autotrophs, using sunlight to convert carbon dioxide and water into sugars by ◊photosynthesis.

The total ◊biomass of autotrophs is far greater than that of animals, reflecting the dependence of animals on plants, and the ultimate dependence of all life on energy from the Sun—green plants convert light energy into a form of chemical energy (food) that animals can exploit. Some bacteria use the chemical energy of sulphur compounds to synthesize organic substances. See also ◊heterotroph.

auxin a plant ◊hormone that promotes stem and root growth in plants. Auxins influence many aspects of plant growth and development, including cell enlargement, inhibition of development of axillary buds, ◊tropisms, and the initiation of roots. ***Synthetic auxins*** are used in rooting powders for cuttings, and in some weedkillers, where high auxin concentrations cause such rapid growth that the plants die. They are also used to prevent premature fruitdrop in orchards. The most common naturally occuring auxin is known as indoleacetic acid (IAA). It is synthesized in the shoot tip and transported to other parts of the plant.

axon the long threadlike extension of a ◊neuron (nerve cell) that conducts electrochemical impulses away from the cell body towards other neurons, or towards an effect or organ such as a muscle. Axons terminate in ◊synapses with other neurons, muscles, or glands.

B

backcross a breeding technique used to determine the genetic make-up (genotype) of an individual organism. To find out whether an organism is ◊homozygous (possessing two dominant alleles) or ◊heterozygous (possessing one dominant and one recessive allele) for a given dominant characteristic observed in its phenotype, it is crossed with another organism that displays the recessive characteristic and is therefore homozygous (possessing two recessive alleles). If all the offspring of that union display the dominant characteristic then the test organism is homozygous; however, if roughly half of the off-spring show the recessive characteristic, the test organism is heterozygous.

bacteria (singular ***bacterium***) microscopic unicellular organisms with prokaryotic cells (see ◊prokaryote). They usually reproduce by ◊binary fission, and since this may occur approximately every 20 minutes, a single bacterium is potentially capable of producing 16 million copies of itself in a day.

Some bacteria are parasitic, causing disease; others, if not checked by refrigeration or other preservation measures, can spoil food and cause food poisoning. However, most bacteria perform essential roles in the ecosystem—digesting ◊cellulose in the intestines of herbivores, bringing about the decomposition of waste or dead matter, and improving soil fertility by processing nitrogen compounds (see ◊nitrogen cycle). Others perform useful functions in the dairy industry—fermenting cheeses and yoghurt—and in the treatment of sewage.

Bacteria are classified biochemically, but their varying shapes provide a rough classification, for example, ***cocci*** are round or oval, ***rods*** are cylindrical, and ***spirillae*** are spiral.

balanced diet diet that includes carbohydrate, protein, fat, vitamins, water, minerals, and roughage. Although it is agreed that all these substances are needed if a person is to be healthy, argument occurs over how much of each type a person needs.

ball-and-socket-joint a ◊joint allowing considerable free movement in three dimensions, for instance the joint between the pelvis and the femur. To facilitate movement, such joints are lubricated by ◊cartilage and synovial fluid. The bones are kept in place by ligaments and moved by muscles.

bark protective outer layer on the stems and roots of woody plants, composed mainly of dead cells. To allow for expansion of the stem, the bark is continually added to from within, and the outer surface often becomes cracked or is shed as scales. Trees deposit a variety of chemicals in their bark, including poisons. Many of these chemical substances have economic value because they can be used in the manufacture of drugs. Quinine, derived from the bark of the *Cinchona* tree, is used to fight malarial infections; curare, an anaesthetic used in medicine, comes from the *Strychnus toxifera* tree in the Amazonian rainforest.

Bark technically includes all the tissues external to the vascular ◊cambium (the ◊phloem, cortex, and periderm), and its thickness may vary from 2.5 mm to 30 cm or more, as in the giant redwood *Sequoia* where it forms a thick, spongy layer.

basal metabolic rate (BMR) the amount of energy needed by an animal just to stay alive. It is measured when the animal is awake but resting, and includes the energy required to keep the heart beating, sustain breathing, repair tissues, and keep the brain and nerves functioning. Measuring the animal's consumption of oxygen gives an accurate value for BMR, because oxygen is needed to release energy from food.

base pair the linkage of two base molecules in ◊DNA. One base lies on one strand of the DNA double helix, and one on the other, so that the base pairs link the two strands like the rungs of a ladder. In DNA, there are four bases: adenine and guanine (purines) and cytosine and thymine (pyrimidines). Adenine always pairs with thymine, and cytosine with guanine. The pairing forms the basis of the ◊genetic code and of heredity.

Because the pairing is so specific, DNA can replicate itself precisely, by separating the two strands of the helix and forming new strands using the original ones as templates. When ◊RNA is formed from DNA, base pairs are again used to copy the genetic message, but in RNA another base, uracil, substitutes for thymine.

behaviour observable activity carried out by an animal in response to a stimulus from its external or internal environments. For example, a monkey may let out a warning cry on glimpsing a bird of prey (external stimulus), or begin to search for food on feeling hungry (internal stimulus).

Behaviour can vary between members of the same species (for example, reproductive behaviour will vary between male and female) or at different times of year. See also ◊social behaviour, ◊instinct, and ◊learning.

beriberi inflammation of the nerve endings, mostly occurring in the tropics and resulting from a deficiency of vitamin B_1 (thiamine).

berry a fleshy, many-seeded ◊fruit that does not split open to release the seeds. The fruit wall, or pericarp, may be juicy and nutritious, and possess a brightly coloured outer skin in order to attract animals to eat the fruit and thus disperse the seeds. Examples of berries are the tomato and the grape.

bicarbonate indicator pH indicator sensitive enough to show a colour change as the concentration of the gas carbon dioxide increases. The indicator is used in photosynthesis and respiration experiments to find out whether carbon dioxide is being liberated. The initial red colour changes to yellow as the pH becomes more acidic.

Carbon dioxide, even in the concentrations found in exhaled air, will dissolve in the indicator to form a weak solution of carbonic acid, which will lower the pH and therefore give the characteristic colour change.

bicuspid valve in the left side of the ◊heart, a flap of tissue that prevents blood from flowing back into the atrium when the ventricle contracts.

biennial plant a plant that completes its life cycle in two years. During the first year it grows vegetatively and the surplus food produced is stored in its ◊perennating organ, usually the root. In the following year these food reserves are used for the production of leaves, flowers and seeds, after which the plant dies. Many root vegetables are biennials, including the carrot and the parsnip.

bile a brownish fluid produced by the liver. It is stored in the gall bladder and emptied into the small intestine as food passes through. Bile consists of bile salts, bile pigments, cholestrol, and lecithin. ***Bile salts*** assist in the breakdown and absorption of fats; ***bile pigments*** are the breakdown products of old red blood cells that are passed into the gut to be eliminated with the faeces.

binary fission form of ◊asexual reproduction, whereby a single-celled organism divides into two smaller 'daughter' cells. It can also occur in a few simple multicellular organisms, such as sea anemones, producing two smaller sea anemones of equal size.

binocular vision or ***stereoscopic vision*** vision in which the eyes face forwards and can both focus on an object at the same time. It is characteristic of predatory animals, including humans.

Although the eyes provide two slightly different images of the world, these are coordinated by the brain to give a three-dimensional perception that allows the animal to judge accurately the position and speed of prey. See also ◊monocular vision.

binomial system of nomenclature the system in which all organisms are identified by a two-part Latinized name. The first name is capitalized and identifies the ◊genus; the second identifies the ◊species within that genus.

Usually the names are descriptive. Thus, the name of the dog, *Canis familiaris*, means the 'familiar species of the dog genus', *Canis* being Latin for 'dog'. See also ◊classification.

biochemistry science concerned with the chemistry of living organisms: the structure and reactions of proteins such as enzymes, nucleic acids, carbohydrates, and lipids.

The study of biochemistry has increased our knowledge of how animals and plants react with their environment, for example, in creating and storing energy by photosynthesis, taking in food and releasing waste products, and passing on their characteristics through their genes. It plays a part in many areas of research, including medicine and agriculture.

biodegradable capable of being broken down by living organisms, principally bacteria and fungi. Biodegradable substances, such as food and sewage, can therefore be rendered harmless by natural processes. The process of decay leads to the release of nutrients that are then recycled by the ecosystem. Nonbiodegradable substances, such as glass, heavy metals, and most types of plastic, present major problems of disposal. See ◊decomposer.

biofeedback modification or control of a biological system by its results or effects.

Many biological systems are controlled by ***negative feedback***. When enough of the hormone thyroxin has been released into the blood, the hormone adjusts its own level by 'switching off' the gland that produces it. In ecology, as the numbers in a species rise, the food supply available to each individual is reduced. This acts to reduce the population to a sustainable level.

biogeography the study of how and why plants and animals are distributed around the world, in the past as well as in the present.

biological control the control of pests such as insects and fungi through biological means, rather than the use of chemicals. This can include breeding resistant crop strains; inducing sterility in the pest; infecting the pest species with disease organisms; or introducing the pest's natural predator. Biological control tends to be naturally self-regulating, but as ecosystems are so complex, it is difficult to predict all the consequences of introducing a biological controlling agent.

biology the study of living things. There are many specializations, including cytology (the study of cells); zoology (the study of animals); and ecology (the study of the environment).

biomass the total mass of living organisms present in a given area. It may be specified for a particular species (such as earthworm biomass) or for a general category (such as herbivore biomass). Estimates also exist for the entire global plant biomass. Measurements of biomass can be used to study interactions between organisms, the stability of those interactions, and variations in population numbers.

biome a broad natural assemblage of plants and animals shaped by common patterns of vegetation and climate. Examples include the tundra biome and the desert biome.

biosphere or ***ecosphere*** the narrow zone that supports life on our planet. It is limited to the waters of the Earth, a fraction of its crust, and the lower regions of the atmosphere.

biosynthesis the synthesis of organic chemicals from simple inorganic ones by living cells—for example, the conversion of carbon dioxide and water to glucose by plants during ◊photosynthesis. Other biosynthetic reactions produce cell constituents including proteins and fats.

Biosynthesis requires energy; in the initial stages of photosynthesis this is obtained from sunlight, but more often it is supplied by the ◊ATP molecule.

biotechnology the industrial use of living organisms to manufacture food, drugs, or other products. The brewing and baking industries have long relied on the yeast microorganism for ◊fermentation purposes, while the dairy industry employs a range of bacteria and fungi to convert milk into cheeses and yoghurts. Recent advances include ◊genetic engineering, in which single-celled organisms with modified ◊DNA are used to produce insulin and other drugs. ◊Enzymes, whether extracted from cells or produced artificially, are central to most biotechnological applications.

biotic factor living variable affecting an ◊ecosystem, for example the changing population of elephants and its effect on the African savanna.

bird member of the vertebrate class Aves, the biggest group of land vertebrates, characterized by warm blood, feathers, forelimbs modified as wings, breathing through lungs, and egg-laying by the female.

birth the act of producing live young from within the body of female animals. Both viviparous and ovoviviparous animals give birth to young. In viviparous animals, embryos obtain nourishment from the mother via a ◊placenta or other means. In ovoviviparous animals, fertilized eggs develop and hatch in the oviduct of the mother and gain little or no nourishment from maternal tissues. See also ◊pregnancy.

birth control another name for ◊family planning.

birth rate or ***natality*** the number of births in a population per unit time; it is usually expressed as the number per year per thousand of the population.

In the 20th century, the UK's birth rate has fallen from 28 to 14 due to increased use of contraception, better living standards and falling infant mortality. The average household now contains 1.8 children. The population growth rate remains high in developing countries. While it is now below replacement level in the UK, in Bangladesh it stands at 33, in Nigeria at 34, and in Brazil at 23 per thousand people per year.

bladder hollow elastic-walled organ in which urine is stored. Urine enters the bladder through two ureters, one leading from each kidney, and leaves it through the urethra.

blind spot area where the optic nerve and blood vessels pass through the retina of the ◊eye. No visual image can be formed as there are no light-sensitive cells in this part of the retina.

blood transport liquid circulating in the arteries, veins, and capillaries of vertebrate animals. It carries nutrients and oxygen to individual cells and removes waste products, such as carbon dioxide. It also plays an important role in the immune response (see ◊immunity) and, in many animals, in the distribution of heat throughout the body.

In humans it makes up 5% of the body weight, occupying a volume of 5.5 litres in the average adult. It consists of a colourless, transparent liquid called ***plasma***, containing microscopic cells of three main varieties. ***Red cells*** (erythrocytes) form nearly half the volume of the blood, with 5 billion cells per litre. Their red colour is caused by ◊haemoglobin. ***White cells*** (leucocytes) are of various kinds. Some (◊phagocytes) ingest invading bacteria and so protect the body from disease; these also help to repair injured tissues. Others (◊lymphocytes) produce antibodies, which help provide immunity. Cell fragments called ***platelets*** (thrombocytes) assist in ◊blood clotting.

Blood cells constantly wear out and die, and are replaced from the bone marrow. Dissolved in the plasma are salts, proteins, sugars, fats, hormones, and fibrinogen, which are transported around the body, the last having a role in clotting.

blood

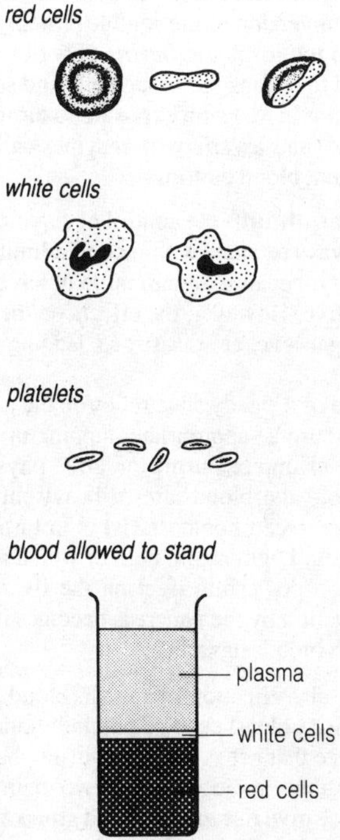

blood clotting a complex series of events that prevents excessive bleeding and the entry of disease-causing microorganisms after injury.

It results in the formation of a meshwork of protein fibres and blood cells over the wound.

When platelets (cell fragments) in the bloodstream come into contact with a damaged blood vessel, they and the vessel wall itself release the enzyme ***thrombokinase***, which brings about the conversion of the inactive enzyme ***prothrombin*** into the active ***thrombin***. Thrombin in turn catalyses the conversion of the soluble protein ***fibrinogen***, present in blood plasma, to the insoluble ***fibrin***. This fibrous protein forms a net over the wound that traps red blood cells and seals the wound; the resulting jellylike clot hardens on exposure to air to form a hard scab. Calcium, vitamin K, and a variety of enzymes called factors are also necessary for efficient blood clotting.

blood glucose regulation the control of sugar concentration in the blood, so that it always remains within certain limits. Blood sugar levels are likely to vary, because an animal may have just eaten or have been physically active. However, the effects on the nervous system of an erratic blood sugar level are disastrous, leading in humans to coma or fits.

The maintenance of a steady state relies on the ability of the liver to store or release glucose as appropriate, topping up blood sugar levels when needed (for example, during or after physical exercise) and removing sugar from the blood after a heavy meal. This important homeostatic function (see ◊homeostasis) is in turn controlled by two hormones, ◊insulin and ◊glucagon, both of which are produced by the pancreas and have a powerful effect on the liver. ◊Diabetes occurs when insulin production by the pancreas becomes insufficient, leading to dangerously high blood sugar levels.

blood group the classification of human blood types according to antigenic activity. Red blood cells of one individual may carry molecules on their surface that act as ◊antigens in another individual whose red blood cells lack these molecules. The two main antigens are designated A and B. These give rise to four blood groups: having A only (A), having B only (B), having both (AB), and having neither (O). Each of these groups may or may not contain the ◊rhesus factor. Correct typing of blood groups is vital in transfusion, since incompatible types of donor and recipient blood will result in blood clotting, with possible death of the recipient.

blood vessel specialist tube that carries blood around the body of multicellular animals. Blood vessels are highly evolved in vertebrates where the three main types, the ◊arteries (and the smaller arterioles), ◊veins (and the smaller venules), and ◊capillaries, are all adapted for their particular role within the body.

blubber thick layer of ◊fat under the skin of marine mammals (seals and whales); it provides an energy store and an effective insulating layer, preventing the loss of body heat to the surrounding water.

BMR abbreviation for ◊***basal metabolic rate*** .

bolus a mouthful of chewed food mixed with saliva, ready for swallowing. Most vertebrates swallow food immediately, but grazing mammals chew their food a great deal, allowing a mechanical and chemical breakdown to begin.

bone the hard skeletal tissue of most vertebrate animals, important in support, movement, and protection. It consists of a network of protein fibres impregnated with minerals, a combination that gives the bone great strength, comparable in some cases with that of reinforced concrete. There are two types of bone: those that develop in the embryo by replacing ◊cartilage and those that form directly from connective tissue. The latter are usually platelike in shape, and include the bones of the cranium. Humans have about 206 distinct bones in the ◊skeleton, of which the smallest are the three ossicles in the middle ear. The interior of the long bones of the limbs comprises a spongy matrix filled with soft marrow, which produces blood cells.

bone marrow soft tissue in the centre of some limb bones that manufactures red and white blood cells.

botany the study of plants. Horticulture, agriculture, and forestry are specialized branches of botany.

Bowman's capsule in the kidney, a microscopic filtering device used in the initial stages of waste-removal and urine formation. There are approximately a million of these capsules in a human kidney, each

bone

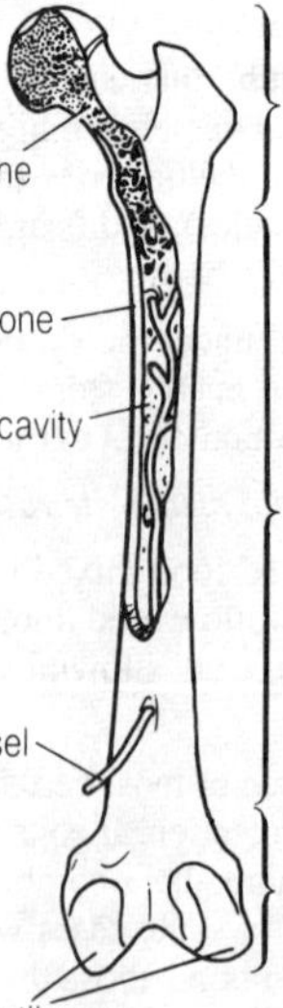

made up of a tight knot of capillaries and each leading into a kidney tubule or nephron. Blood at high pressure passes into the capillaries where water, dissolved nutrients, and urea move through the capillary wall into the tubule.

brain in higher animals, a mass of interconnected ◊neurons (nerve cells), forming the anterior part of the ◊central nervous system, whose activities it coordinates and controls. In ◊vertebrates, the brain is contained by the skull.

It is composed of three main regions. An enlarged portion of the upper spinal cord, the ***medulla***, contains centres for the control of involuntary processes such as respiration, heartbeat rate and strength, and blood pressure. Overlying this is the ***cerebellum***, which is concerned with coordinating complex muscular processes such as maintaining posture and moving limbs. The cerebral hemispheres (***cerebrum***) are paired outgrowths of the front end of the brain, involved, in the higher vertebrates, in the integration of all sensory input and motor output, in intelligent behaviour, and, in humans, in language.

breast one of a pair of ◊mammary glands on the upper front of the human female. Each of the two breasts contains milk-producing cells, and a network of tubes or ducts that lead to an opening in the nipple. Milk-producing cells in the breast do not become active until a woman has given birth to a baby. Breast milk is made from substances extracted from the mother's blood as it passes through the breasts. It contains all the nourishment a baby needs, including antibodies to help fight infection.

breathing in land animals, the muscular movements by which air is taken into the lungs and then expelled, a form of ◊gas exchange.

breathing

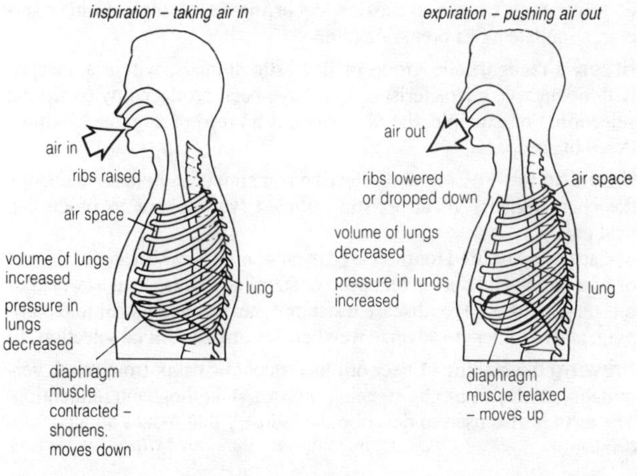

Breathing is sometimes referred to as external respiration because true respiration is a cellular (internal) process.

Lungs are specialized ◊respiratory surfaces but are not themselves muscular, consisting of spongy material. In order for oxygen to be passed to the blood and carbon dioxide removed, air is forced in and out of the chest region by the ribs and accompanying intercostal muscles, the rate of breathing being controlled by the brain. High levels of activity lead to a greater demand for oxygen and a subsequent higher rate of breathing.

breathing rate the number of times a minute the lungs inhale and exhale. The rate increases during exercise because the muscles require an increased supply of oxygen and nutrients. At the same time very active muscles produce a greater volume of carbon dioxide, a waste gas that must be removed by the lungs via the blood.

The regulation of the breathing rate is under both voluntary and involuntary control, although a person can only forcibly stop breathing for a limited time. The regulatory system includes the use of chemoreceptors, which can detect levels of carbon dioxide in the blood. High concentrations of carbon dioxide, occurring for example during exercise, stimulate a fast breathing rate.

breed a recognizable group of domestic animals, within a species, with distinctive characteristics that have been produced by ◊artificial selection. For example, the St Bernard is a breed of dog, the Merino a breed of sheep.

breeding the crossing and selection of animals and plants to change the characteristics of an existing ◊breed (variety), or to produce a new one.

Cattle may be bred for increased meat or milk yield, sheep for thicker or finer wool, and horses for speed or stamina. Plants, such as wheat or maize, may be bred for disease resistance, heavier and more rapid cropping, and hardiness to adverse weather. See also ◊artificial selection.

brewing the making of beer or other alcoholic drink from dried, germinating barley (malt) by steeping (mashing), boiling, and fermenting. The term is also used to describe the industry that makes all alcoholic drinks.

Mashing releases the ◊maltose sugar naturally present in germinating barley. Yeast is then added, which respires anaerobically, breaking down the maltose to give ethanol (alcohol) and carbon dioxide. Hops are added to give a bitter taste. See ◊anaerobic respiration.

bronchiole narrow air tube found in the ◊lung responsible for delivering air to the main respiratory surfaces. Bronchioles lead off from the larger ◊bronchus and branch extensively before terminating in the many thousands of alveoli that form the bulk of the lung tissue.

bronchus (plural ***bronchi***) one of a pair of large tubes splitting off from the trachea (windpipe) and passing into the lung. Apart from their size, bronchi differ from the ◊bronchioles in possessing cartilaginous rings, which give rigidity and prevent collapse during breathing movements.

Numerous glands secrete a slimy mucus, which traps dust and other particles, and is constantly being propelled upwards to the mouth by thousands of tiny hairs or cilia. The bronchus is adversely affected by several respiratory diseases and by smoking, which damages the cilia and therefore the lung cleaning mechanism.

bud an undeveloped shoot usually enclosed by protective scales; inside is a very short stem and numerous undeveloped leaves, or flower parts, or both. Terminal buds are found at the tips of shoots, while axillary buds develop in the axils (angles between stem and leaf), often remaining dormant unless the terminal bud is removed or damaged.

budding a type of ◊asexual reproduction in which an outgrowth develops from a cell to form a new individual. Most yeasts reproduce in this way.

In a suitable environment, yeasts grow rapidly, forming long chains of cells as the buds themselves produce further buds before being separated from the parent. Simple invertebrates, such as ◊hydra, can also reproduce by budding.

bulb underground bud with fleshy leaves containing a reserve food supply and with roots growing from its base. Bulbs function in vegetative reproduction and are characteristic of many monocotyledenous plants such as the daffodil, snowdrop, and onion.

bur or ***burr*** type of 'false fruit' or ◊pseudocarp covered with many hooks; for instance, that of burdock. The term is also used to include any type of fruit or seed bearing hooks, such as that of goosegrass. Burs catch in the feathers or fur of passing animals, and thus may be dispersed (see ◊dispersal) over considerable distances.

C

caecum in the ◊digestive system, a blind-ending tube branching off from the first part of the large intestine, terminating in the appendix. It is used for the digestion of cellulose by some grass-eating mammals; the rabbit caecum and appendix, for example, contain millions of bacteria that produce cellulase, the enzyme necessary for the breakdown of cellulose to glucose. In humans, the caecum and appendix are ◊vestigial organs, having no function.

calorific value the amount of heat energy generated by a given mass of food when it is completely burned, or when it is completely oxidized by aerobic respiration. It is measured in joules (or kilojoules).

The calorific value of a food item can be measured experimentally by burning that item and using the heat released to raise the temperature of a known mass of water.

calorific value (J) = mass of water (g) × temperature rise (°C) × 4.2

calyx the collective term for the ◊sepals of a flower, forming the outermost whorl of the ◊perianth. It surrounds the other flower parts and protects them while in bud.

cambium a layer of actively dividing cells (lateral ◊meristem), found within stems and roots, that gives rise to ◊secondary growth in perennial plants, causing an increase in girth. There are two main types of cambium: vascular cambium, which gives rise to secondary ◊xylem and ◊phloem tissues, and cork cambium, which gives rise to secondary cortex tissues (see ◊bark).

cancer group of diseases characterized by abnormal proliferation of cells. Cancer (malignant) cells are usually degenerate, capable only of reproducing themselves (forming tumours) until they outnumber the surrounding healthy cells. Malignant cells tend to spread from their site of origin by travelling through the bloodstream or lymphatic system.

There are more than 100 types of cancer. Some, like lung or bowel cancer, are common; others are rare. The likely cause remains unexplained. Triggering agents (carcinogens) include chemicals such as those found in cigarette smoke, other forms of smoke, asbestos dust, exhaust fumes, and many industrial chemicals. Some viruses can also trigger the cancerous growth of cells, as can X-rays and radioactivity. Dietary factors are important in some cancers; for example, lack of fibre in the diet may predispose people to bowel cancer.

canine in mammalian carnivores, long, often pointed teeth found at the front of the mouth between the incisors and premolars. They are used for catching prey, for killing, and for tearing flesh. Canines are absent in herbivores such as rabbits and the sheep, and are much reduced in humans.

capillarity or ***capillary action*** the spontaneous movement of liquids up or down narrow tubes such as plant xylem vessels (see ◊transpiration) or the spaces existing between ◊soil particles.

capillary the narrowest blood vessel in vertebrates, between 8 and 20 micrometres in diameter, barely wider than a red blood cell. Capillaries are distributed as ***beds***, complex networks connecting arteries and veins. Capillary walls are extremely thin, consisting of a single layer of cells, and so nutrients, dissolved gases, and waste products can easily pass through them. This makes the capillaries the main area of exchange between the fluid (◊lymph) bathing body tissues and the blood.

capsule dry, usually many-seeded fruit formed from an ovary composed of two or more fused ◊carpels, which bursts open to disperse its seeds.

The seeds may be released from the capsule through pores, as in the poppy; through a lid, as in the scarlet pimpernel; or through lengthwise splits, as in the crocus.

carbohydrate organic molecule containing carbon, hydrogen, and oxygen, and possessing the general formula $Cx(H_2O)y$.

The basic unit of all carbohydrates is a simple sugar molecule (monosaccharide), such as glucose. These units may be combined into disaccharides (two sugars), such as sucrose, or into polysaccharides (many sugars linked together in a chain), such as starch, glycogen, and cellulose.

Carbohydrates are an important part of a balanced human diet, providing energy for life processes including growth and movement. Excess carbohydrate intake can be converted into fat and stored in the body.

carbon cycle the sequence by which carbon circulates and is recycled through the natural world. The carbon element from carbon dioxide, released into the atmosphere by animals as a result of ◊respiration, is taken up by plants during ◊photosynthesis and converted into carbohydrates; the oxygen component is released back into the atmosphere. The simplest link in the carbon cycle, however, occurs when an

carbon cycle

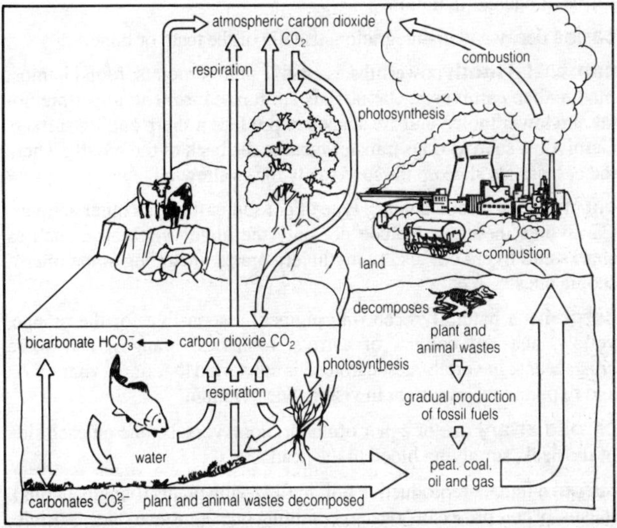

animal eats a plant and carbon is transferred from, say, a leaf cell to the animal body. Today, the carbon cycle is being disrupted by the increased consumption and burning of fossil fuels, and the burning of large tracts of tropical forests, as a result of which levels of carbon dioxide are building up in the atmosphere and probably contributing to the ◊greenhouse effect.

carbon dioxide CO_2 colourless gas produced by living things during ◊respiration and by the decomposition or burning of organic matter. Green plants need carbon dioxide for ◊photosynthesis.

It accounts for less than 0.04% of the gases in the atmosphere; however, its increasing density due to the burning of fossil fuels is contributing to the ◊greenhouse effect and global warming.

carcinogen any agent that increases the chance of a cell becoming cancerous, including various chemicals, some viruses, X-rays, and forms of nuclear radiation.

caries decay and disintegration, usually of the teeth or bone.

carnassial tooth powerful scissor-like pair of molars, found in most mammalian carnivores. Carnassials are formed from an upper premolar and lower molar, and are shaped to produce a sharp cutting surface. Carnivores such as dogs transfer meat to the back of the mouth, where the carnassials slice up the food ready for swallowing.

carnivore in a food chain, a type of consumer that eats other animals. Carnivores therefore include not only the large vertebrates, such as sharks and tigers but also many invertebrates, including some microscopic ones.

carotene a naturally occurring pigment responsible for the orange, yellow, and red colours of carrots, tomatoes, oranges, and some crustaceans. In vertebrates, carotene is converted to retinol (vitamin A) and to pigments important in vision. See ◊vitamin.

carotid artery one of a pair of major blood vessels, one on each side of the neck, supplying blood to the head.

carpel a female reproductive unit in flowering plants (◊angiosperms). It usually comprises an ◊ovary containing one or more ovules, a ◊style, and a ◊stigma, which receives the pollen. A flower may have one or more carpels, and they may be separate or fused together. Collectively the carpels of a flower are known as the ◊gynoecium.

carrying capacity in ecology, the maximum number of animals of a given species that a particular area can support. When the carrying capacity is exceeded, there is insufficient food (or other resources) for the members of the population. The population may then be reduced by emigration, reproductive failure, or death through starvation.

cartilage flexible bluish-white ◊connective tissue made up of the protein collagen. In cartilaginous fish it forms the skeleton; in other vertebrates it forms the greater part of the the embryonic skeleton, and is replaced by ◊bone in the course of development, except in areas of wear such as bone endings, and the discs between the backbones. It also forms structural tissue in the larynx, nose, and external ear (pinna) of mammals.

casein main protein of milk, from which it can be separated by the action of acid, the enzyme rennin, or bacteria (souring); it is also the main component of cheese.

catabolism the destructive part of ◊metabolism in which living tissue is changed into energy and waste products. It is the opposite of ◊anabolism. Catabolism occurs continuously in the body, but is accelerated during fevers and in starvation.

catalyst substance that alters the speed of (or makes possible) a chemical or biological reaction but that remains unchanged at the end of the reaction. ◊Enzymes are natural biological catalysts.

catkin in flowering plants (angiosperms), a hanging inflorescence, bearing numerous small, usually unisexual flowers. Many types of trees bear catkins, including willows, poplars, and birches. Most plants with catkins are wind-pollinated, so the male catkins produce large quantities of pollen.

cell discrete, membrane-bound portion of living matter, the smallest unit capable of an independent existence. Each is formed by the cell division (◊mitosis or ◊meiosis) of another cell. All living organisms consist of one or more cells, with the exception of viruses. Bacteria, protozoa, and many other microorganisms consist of single cells, whereas a human is made up of billions of cells.

Essential features of a cell are the membrane, which encloses it and restricts the flow of substances in and out; the jellylike material, or ◊cytoplasm, within; the ◊ribosomes, which carry out protein synthesis: and the ◊DNA, which forms the hereditary material. The cells of plants, bacteria, and fungi possess a rigid outer cell wall that protects the cell and maintains its shape.

In the ◊eukaryotes (protozoa, fungi, and higher animals and plants), DNA is organized into ◊chromosomes and contained within a ◊nucleus. The only cells of the human body that have no nucleus are the red blood cells. The nuclei of some cells contain a denser spot called the ◊nucleolus. The eukaryotic cell also possesses organelles such as ◊mitochondria, ◊chloroplasts, ◊endoplasmic reticulum, ◊Golgi apparatus and ◊centrioles, which perform specialized tasks.

In ◊prokaryotes (bacteria and cyanobacteria), the DNA forms a simple loop and there is no nucleus.

cell division the process by which a cell divides, either ◊meiosis, associated with sexual reproduction, or ◊mitosis, associated with growth, cell replacement or repair. Both forms involve the duplication of DNA and the splitting of the nucleus.

cell membrane or ***plasma membrane*** thin semipermeable layer of protein and fat that encloses cells and controls the flow substances into and out of the cytoplasm. Generally, small molecules such as water, glucose, and amino acids can penetrate the membrane, while large molecules such as starch cannot. Membranes also play a part in ◊active transport, hormonal response, and cell metabolism.

cell sap dilute fluid found in the large central vacuole of many plant cells. It is made up of water, amino acids, glucose, and salts. The sap has many functions, including storage of useful materials, and provides mechanical support for non-woody plants.

cellulose ◊polysaccharide composed of long chains of glucose units. It is the most abundant substance found in the plant kingdom, being the principal constituent of the cell wall of higher plants, and is a vital ingredient in the diet of many ◊herbivores. Molecules of cellulose are organized into long, unbranched microfibrils that give support to the cell wall.

No vertebrate is able to produce cellulase, the enzyme necessary for the breakdown of cellulose into sugar. Yet most mammalian herbivores (such as rabbits and cows) rely on cellulose, using secretions from microorganisms living in the gut to break it down. Humans cannot digest cellulose; they possess neither the correct gut microorganisms nor the necessary grinding teeth. However, cellulose still forms a necessary part of the human diet as ◊roughage.

cell wall in plants, bacteria, and fungi, the tough outer surface of the cell. The cell wall of plants is constructed from a mesh of ◊cellulose and is very strong and relatively inelastic. Most living plant cells are turgid (swollen with water; see ◊turgor) and develop an internal hydrostatic pressure (wall pressure) that acts against the cellulose wall. The result of this turgor pressure is to give the cell, and therefore the plant, rigidity. Plants that are not woody are particularly reliant on this form of support.

central nervous system the part of the nervous system with a concentration of ◊neurons (nerve cells) that coordinates body functions. In ◊vertebrates, the central nervous system consists of a brain and a dorsal nerve cord (the spinal cord) within the spinal column.

centriole in animals, a structure found in the ◊cells that plays a role in the processes of ◊meiosis and ◊mitosis (cell division).

centromere part of the ◊chromosome where there are no ◊genes. Under the microscope, it usually appears as a constriction in the strand of the chromosome, and is the point at which the spindle fibres are attached during ◊meiosis and ◊mitosis (cell division).

cereal grass grown for its edible starch seeds. The term refers primarily to barley and wheat, but may also refer to oats, maize, rye, millet, and rice. It stores easily, and contains about 75% carbohydrate and 10% protein. If all the world's cereal crop were consumed directly by humans, everyone could obtain adequate protein and carbohydrate; however, a large proportion of cereal production, especially in affluent nations, is used as animal feed to boost the production of meat, milk, butter, and eggs.

cerebellum in vertebrates, the part of the brain that controls muscular movements, balance, and coordination. The human cerebellum is well developed, because of the need for balance when walking or running, and for coordinated hand movements.

cerebral hemisphere one of the two halves of the ◊cerebrum.

cerebrum part of the vertebrate ◊brain, formed from two paired cerebral hemispheres. In birds and mammals it is the largest part of the brain. It is covered with an infolded layer of grey matter, the ***cerebral cortex***, which integrates brain functions. The cerebrum coordinates the senses, and is responsible for learning and other higher mental faculties.

cervix (Latin 'neck') the neck of the ◊uterus; a narrow passage that opens into the vagina.

chemosynthesis process by which certain bacteria obtain energy from chemical reactions and use it to drive the processes that synthesize the organic compounds they require. Unlike ◊photosynthesis, it does not require the presence of light.

For example, nitrogen-fixing bacteria are a group of chemosynthetic organisms that change free nitrogen into a form that can be taken up by plants (see ◊nitrogen cycle).

chemotropism movement by part of a plant in response to a chemical stimulus. The response by the plant is termed 'positive' if the growth is towards the stimulus or 'negative' if the growth is away from the stimulus.

Fertilization of flowers by pollen is achieved because the ovary releases chemicals that produce a positive chemotropic response from the developing pollen tube.

childbirth the expulsion of a baby from its mother's body following ◊pregnancy.

chlorofluorocarbon (CFC) nontoxic odourless chemical, used as a propellant in aerosol cans, as a refrigerant in refrigerators and air conditioners, and in the manufacture of foam boxes for take-away food cartons. CFCs are partly responsible for the destruction of the ozone layer.

When CFCs drift up into the upper atmosphere they break down, under the influence of the Sun's ultraviolet radiation, into chlorine atoms, which destroy the ozone layer and allow harmful radiation from the Sun to reach the Earth's surface. CFCs can remain in the atmosphere for more than 100 years. Production of CFCs is now being phased out in most countries, and replacements developed.

chlorophyll green pigment present in most plants that is responsible for the absorption of light energy during ◊photosynthesis. The pigment absorbs the red and blue-violet parts of sunlight but reflects the green, thus giving plants their most characteristic colour.

Chlorophyll is found within chloroplasts, which are present in large numbers in the ◊palisade cells of leaves. It is similar in structure to ◊haemoglobin, but with magnesium instead of iron as the reactive part of the molecule.

chloroplast a structure (◊organelle) within a plant cell containing the green pigment chlorophyll. Chloroplasts occur in most cells of the green plant that are exposed to light, often in large numbers. Typically, they are flattened and disc-like, with a double membrane enclosing the stroma, a gel-like matrix. Within the stroma are stacks of fluid-containing cavities, or vesicles, where ◊photosynthesis occurs.

chloroplast

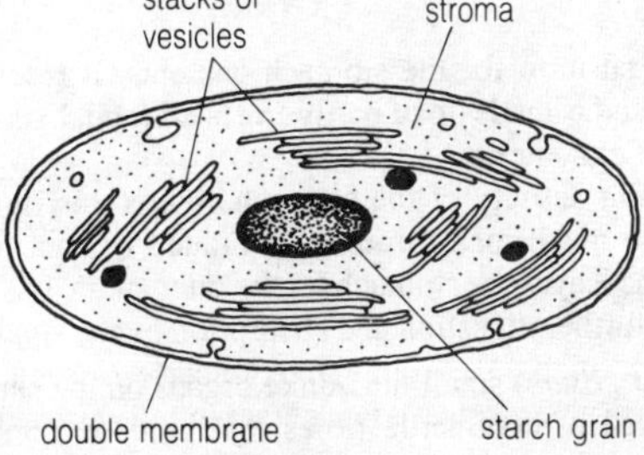

chlorosis an abnormal condition of green plants in which the stems and leaves turn pale green or yellow. The yellowing is due to a reduction in the levels of the green chlorophyll pigments. It may be caused by a deficiency in essential elements (such as magnesium, iron, or manganese), a lack of light, genetic factors, or virus infection.

cholesterol a waxy, fatty substance abundant in the bodies of animals and found also in some plants. It is a vital constituent of cell membranes and is the starting point in the formation of many hormones including the sex hormones. Cholestrol is manufactured in the liver and intestine, and is also provided in the human diet by foods such as eggs, meat, and butter. A high level of cholesterol in the blood is thought to contribute to ◊arteriosclerosis (hardening of the arteries).

choroid the black layer found at the rear of the ◊eye beneath the retina. By absorbing light that has already passed through the retina, it stops back- reflection and so aids vision.

chromosome a structure in a cell nucleus that carries the ◊genes. Each chromosome consists of one very long strand of DNA, coiled and folded to produce a compact chromosome. The point on a chromosome where a particular gene occurs is known as its locus. Most higher organisms have two copies of each chromosome (they are ◊diploid) but some have only one (they are ◊haploid). See also ◊mitosis and ◊meiosis.

chyle thin milky ◊lymph fluid found in the ◊lacteals and containing food materials, such as emulsified fats, absorbed from the small intestine.

chyme general term for the stomach contents. It resembles a thick creamy fluid and is made up of partly-digested food, hydrochloric acid, and a range of enzymes.

The muscular activity of the stomach churns this fluid constantly, continuing the mechanical processes initiated by the mouth. By the time the chyme leaves the stomach for the duodenum, it is a smooth liquid ready for further digestion and absorption by the small intestine.

cilia (singular *cilium*) small threadlike organs on the surface of some cells, composed of contractile fibres that produce rhythmic waving movements. Some single-celled organisms move by means of cilia. In multicellular animals, they keep lubricated surfaces clear of debris. They also move food in the digestive tracts of some invertebrates.

chromosome

the 23 pairs of chromosomes of a normal human male (XY)

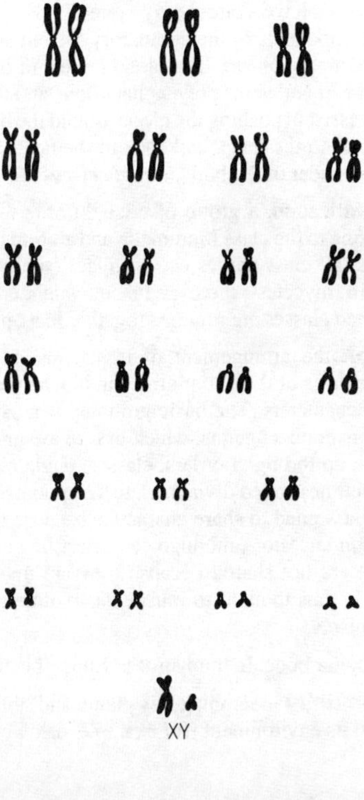

ciliary muscle ring of muscle surrounding and controlling the lens inside the vertebrate eye, used in ◊accommodation (focusing). Suspensory ligaments, resembling the spokes of a wheel, connect the lens to the ciliary muscle and pull the lens into a flatter shape when the muscle relaxes. On contraction, the lens returns to its normal spherical state.

circulatory system in animals, the system of vessels that transports the blood to and from the different parts of the body. Except for simple animals such as sponges and coelenterates (jellyfishes, sea anemones, corals), all animals have a circulatory system.

The main components of the circulatory system are the ◊blood vessels and the pumping ◊heart. The blood travels in one direction only. Valves in the heart and veins prevent backflow, and the muscular walls of the arteries assist in pushing the blood around the body. In mammals, the blood passes to the lungs and back to the heart before circulating around the remainder of the body (***double circulation***).

class in classification, a group of related ◊orders. For example, all mammals belong to the class Mammalia and all birds to the class Aves. Among plants, all class names end in 'idae' (such as Asteridae) and among fungi in 'mycetes'; there are no equivalent conventions among animals. Related classes are grouped together in a ◊phylum.

classification the arrangement of organisms into a hierarchy of groups, on the basis of their similarities in biochemical, anatomical or physiological characters. The basic grouping is a ◊species, several of which may constitute a ◊genus, which in turn are grouped into ◊families, and so on up through ◊orders, classes, phyla (singular ◊phylum; in plants, sometimes called divisions), to ◊kingdoms.

Species are assumed to share characters because they acquire them from a common ancestor (although care must be taken to ensure that the characters are not shared because they are ◊analogous). Such a classification is thus thought to mirror the evolutionary relationships between organisms.

clavicle the collar bone. In humans it is vulnerable to fracture.

climax community assemblage of plants and animals that is relatively stable in its environment (for example, oak woods in Britain). It is brought about by ecological ◊succession, and represents the point at which succession ceases to occur.

clone group of cells or organisms arising by asexual reproduction from a single 'parent' individual. Clones therefore share exactly the same genetic make-up.

Examples include a group of plants produced from cuttings from a single plant.

cochlea part of the inner ◊ear. It is equipped with approximately 10,000 hair cells, which move in response to sound waves and thus stimulate neurons (nerve cells) to send messages to the brain.

codominance in genetics, the failure of a pair of alleles, controlling a particular characteristic, to show the classic recessive-dominant relationship. Instead, aspects of both alleles may show in the phenotype.

The snapdragon shows codominance in respect to colour. Two alleles, one for red petals and the other for white, will produce a pink colour if the alleles occur together as a heterozygous form.

coelom in all but the simplest animals, the fluid-filled cavity that separates the body wall from the gut and associated organs, and allows the gut muscles to contract independently of the rest of the body.

coevolution evolution of structures and behaviours within a species that can best be understood in relation to another species. For example, insects and flowering plants have evolved together: insects have produced mouthparts suitable for collecting pollen or drinking nectar, and plants have developed chemicals and flowers that will attract insects to them.

Coevolution occurs because both groups of organisms, over millions of years, benefit from a continuing association, and will evolve structures and behaviours that maintain this association.

coil another name for the contraceptive ◊intrauterine device.

cold-bloodedness common name for ◊poikilothermy.

coleoptile the protective sheath that surrounds the young shoot tip of a grass during its passage through the soil to the surface. Although of relatively simple structure, most coleoptiles are very sensitive to light, ensuring that seedlings grow upwards. See also ◊auxin.

collagen in vertebrates, a strong, rubbery ◊protein that plays a major structural role. Collagen supports the ear flaps and the tip of the nose in humans, as well as being the main constituent of tendons and ligaments. Bones are made up of collagen, with the mineral calcium phosphate providing increased rigidity.

collenchyma a plant tissue composed of relatively elongated cells with thickened cell walls, in particular at the corners where adjacent cells meet. It is a supporting and strengthening tissue found in non-woody plants, mainly in the stems and leaves.

colon the part of the large intestine between the caecum and rectum, where water and mineral salts are absorbed from digested food, and the residue formed into faeces or faecal pellets for egestion.

colonization the spreading of a species into a new habitat, such as a freshly cleared field, a new motorway verge, or a recently flooded valley. The first species to move in are called ***pioneers***, and may establish conditions that allow other animals and plants to move in (for example, by improving the condition of the soil or by providing shade). Over time a range of species arrives and the habitat matures; early colonizers will probably be replaced, so that the variety of animal and plant life present changes. This is known as ◊succession.

Ecologists can judge the history of a habitat by looking at the organisms present, as certain species of plant and animal are associated with particular stages of colonization and succession.

colourings food ◊additives used to alter or improve the colour of processed foods. They include artificial colours, such as tartrazine and amaranth, which are made from petrochemicals, and 'natural' colours such as chlorophyll, caramel, and carotene. Some of the natural colours are actually synthetic copies of the naturally occurring substances, and some of these, notably the synthetically produced caramels, may be injurious to health.

colour vision the ability of the eye to recognize different frequencies in the visible spectrum as colours. See ◊retina.

commensalism a relationship between two ◊species whereby one (the commensal) benefits from the association, whereas the other neither benefits nor suffers. For example, certain species of millipede and silverfish inhabit the nests of army ants and live by scavenging on the refuse of their hosts, but without affecting the ants.

communication the signalling of information by one organism to another, usually with the intention of altering the recipient's behaviour. Signals used in communication may be ***visual*** (such as the human smile or the display of colourful plumage in birds), ***auditory*** (for example, the whines or barks of a dog), ***olfactory*** (such as the odours released by the scent glands of a deer), ***electrical*** (as in the pulses emitted by electric fish), or ***tactile*** (for example, the nuzzling of male and female elephants).

community in ecology, an assemblage of plants, animals, and other organisms living within a circumscribed area. Communities are usually named by reference to a dominant feature such as characteristic plant species (for example, beech wood community), or a prominent physical feature (for example, a freshwater-pond community).

compensation point the point at which there is just enough light for a plant to survive. At this point all the food produced by ◊photosynthesis is used up by ◊respiration. For aquatic plants, the compensation point is the depth of water at which there is just enough light to sustain life (deeper water = less light = less photosynthesis).

competition interaction between two or more organisms, or groups of organisms (for example, species), that use a common resource that is in short supply. Thus plants may compete with each other for sunlight, or nutrients from the soil, while animals may compete amongst themselves for food, water, or refuges.

Competition invariably results in a reduction in the numbers of one or both competitors, and in ◊evolution contributes both to the decline of certain species and to the evolution of ◊adaptations.

concentration gradient change in the concentration of a substance from one area to another. See ◊diffusion.

conditioning type of learning in which an animal automatically associates a stimulus with an event and modifies its behaviour accordingly. It is believed to involve rather simple processes, and can be easily reproduced and studied in the laboratory. In the original experiments, carried out by the Russian psychologist Ivan Pavlov (1849–1936), a dog was fed normally, but each meal (event) was preceded by the sounding of a bell (stimulus). Soon the dog was conditioned to associate the bell with food, and could be observed to salivate and wag its tail whenever the bell was rung, even before it had sensed the presence of food.

condom or ***sheath*** barrier contraceptive, made of rubber, which fits over an erect penis and holds in the sperm produced by ejaculation. It is an effective means of preventing pregnancy if used carefully, preferably with a ◊spermicide. A condom with spermicide is 97% effective; one without spermicide is 85% effective. Condoms can also give protection against sexually transmitted diseases (STDs).

cone reproductive structure of conifers (gymnosperms), such as pine and fir trees. It consists of a central axis surrounded by overlapping, scale-like leaves, which may be woody. Usually there are separate male and female cones, the former bearing pollen sacs containing pollen grains, and the larger female cones bearing the ovules that contain the ova or egg cells. The pollen is carried from male to female cones by wind. The seeds develop within the female cone and are released as the scales open in dry atmospheric conditions.

cone cell type of cell found in the ◊retina of the eye that plays a part in colour vision.

conjunctiva membrane covering the ◊eye. It is continuous with the epidermis of the eyelids, and lies on the surface of the cornea.

connective tissue in animals, tissue made up of a noncellular substance in which some cells are embedded. Skin, bones, tendons, cartilage, and adipose tissue (fat) are the main connective tissues.

conservation in society, action taken to try and preserve the natural world, in particular attempts to protect it from pollution, destruction, and other harmful features of human activity. The late 1980s saw a great increase in public concern for the environment, with membership of conservation groups, such as Friends of the Earth, rising dramatically. Conservation issues of global significance include the depletion of atmospheric ozone by the action of ◊chlorofluorocarbons (CFCs), the build-up of carbon dioxide in the atmosphere (thought to contribute to an intensification of the ◊greenhouse effect), and the destruction of the tropical rainforests.

consumer any organism that obtains its food by consuming organic material. Animals are consumers, as are the saprophytic fungi, which live off decaying matter.

Herbivores are ***primary consumers***, relying on vegetation. Carnivorous animals are ***secondary consumers***, animals that eat herbivores. ***Tertiary consumers*** eat other carnivores. Secondary and tertiary consumers are less numerous than either the producers or the primary consumers because only about 10% of the energy in any one ◊trophic level can pass into the next level up. See also ◊food chain.

continuous variation slight difference in an individual character, such as height, across a sample of the population. Although there are very tall and very short humans, there are also many people with an intermediate height. The same applies to bodyweight. Continuous variation can result from the genetic make-up of a population, or from environmental influences, or from a combination of the two. Compare ◊discontinuous variation.

contraceptive any drug, device, or technique that prevents pregnancy. The contraceptive pill (the ◊Pill) contains female hormones that interfere with egg production or the first stage of pregnancy. Barrier contraceptives include ◊condoms (sheaths) and ◊diaphragms (caps); they prevent the sperm entering the cervix (neck of the womb). Intrauterine devices (IUDs or coils), cause a slight inflammation of the lining of the womb, which prevents the fertilized egg from becoming implanted. A new development is a sponge impregnated with spermicide that is inserted into the vagina. Other contraceptive methods include sterilization (women) and vasectomy (men); these are usually nonreversible.

'Natural' methods include withdrawal of the penis before ejaculation (coitus interruptus), and avoidance of intercourse at the time of ovulation (◊rhythm method). These methods are unreliable and normally only used on religious grounds. See ◊family planning.

contractile root thickened root at the base of a corm, bulb, or other organ that helps position it at an appropriate level in the ground. After the root has become anchored in the soil, its upper portion contracts, pulling the plant deeper into the ground.

contractile vacuole tiny organelle found in many single-celled fresh-water organisms. Fresh-water protozoa such as *Amoeba* absorb water by the process of ◊osmosis, and this excess must be eliminated. The vacuole expands as it slowly fills with water, and then contracts, expelling the water from the cell.

control the process by which a tissue, an organism, a population, or an ecosystem maintains itself in a balanced, stable state. Control of an organism's internal environment is called ◊homeostasis.

control experiment an essential part of a scientifically valid experiment, designed to show that the factor being tested is actually responsible for the effect observed. In the control experiment all factors, apart from the one under test, are exactly the same as in the test experiments, and all the same measurements are carried out. In drug trials, a placebo (a harmless substance) is given alongside the substance being tested in order to compare effects.

convergent evolution the independent evolution of similar structures in species (or other taxonomic groups) that are not closely related, as a result of living in a similar way. Thus, birds and bats have wings, not because they are descended from a common winged ancestor, but because their respective ancestors independently evolved flight.

In such cases, the structures often differ in their anatomical origins and are only superficially similar. Such structures are said to be ◊analogous, in contrast to the ◊homologous organs of related groups.

copulation the act of mating in animals with internal ◊fertilization. Male mammals have a ◊penis or other organ, which is used to introduce spermatozoa into the reproductive tract of the female.

corm a short, swollen, underground plant stem, surrounded by protective scale leaves, as seen in the crocus. It stores food, provides a means of ◊vegetative reproduction, and acts as a ◊perennating organ.

During the year, the corm gradually withers as the food reserves are used for the production of leafy, flowering shoots formed from axillary buds. Several new corms are formed at the base of these shoots, above the old corm.

cornea transparent front section of the ◊eye. The cornea is curved and behaves as a fixed lens, so that light entering the eye is partly focused before it reaches the lens.

There are no blood vessels in the cornea and it relies on the fluid in the front chamber of the eye for nourishment. Further protection for the eye is provided by the ◊conjunctiva.

cornified layer the upper layer of the skin where the cells have died, their cytoplasm having been replaced by keratin, a fibrous protein also found in nails and hair. Cornification gives the skin its protective water-proof quality.

corolla a collective name for the ◊petals of a flower. In some plants the petal margins are partially or completely fused—for example, in bindweed.

corpus luteum temporary endocrine gland found in the ◊ovary. It is formed after ovulation from the Graafian follicle, a group of cells associated with bringing the egg to maturity, and secretes the hormone progesterone, which maintains the uterus wall ready for pregnancy. If pregnancy does not occur the corpus luteum breaks down.

cortex in animals, the outer layer of a structure such as the brain, kidney, or adrenal gland. In plants, the cortex includes non-specialized cells lying just beneath the surface cells of the root and stem.

cotyledon a structure in the embryo of a seed plant (angiosperm or gymnosperm) that may form the first green 'leaf' after germination and is commonly known as a seed leaf. The number of cotyledons present in an embryo is an important character in the classification of angiosperms. Monocotyledons (such as grasses, palms, and lilies) have a single cotyledon, whereas dicotyledons (the majority of angiosperm species) have two. In gymnosperms there may be up to a dozen cotyledons within each seed.

In seeds that also contain ◊endosperm (nutritive tissue), the

cortex

plant cortex cells

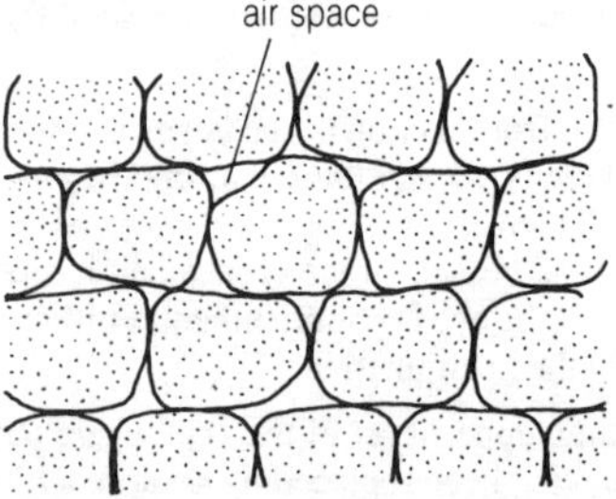

cotyledons are thin, but where they are the primary food-storing tissue, as in peas and beans, they may be quite large.

courtship behaviour exhibited by animals as a prelude to mating. The behaviour patterns vary considerably from one species to another, but are often ritualized forms of behaviour not obviously related to courtship or mating (for example, courtship feeding in birds).

Courtship ensures that copulation occurs with a member of the opposite sex of the right species. It also synchronizes the partners' readiness to mate and allows each partner to assess the suitability of the other. In pigeons, for example, males may reject females that respond too quickly to their courtship displays because such females have probably already mated with another male.

cranium the dome-shaped area of the skull, consisting of several fused plates, that protects the brain.

creationism a theory concerned with the origins of matter and life, claiming, as does the Bible in Genesis, that the world and humanity were created by a supernatural Creator. It conflicts with the theory of ◊evolution.

crop rotation the system of regularly changing the types of crop grown on a piece of land. The crops are grown in a particular order in order to use and replace the nutrients in the soil, and to prevent the build-up of insect and fungal pests.

The crops rotated frequently include a legume, such as a bean or pea species, because the activities of the nitrogen-fixing bacteria in its roots add to the nitrate content of the soil (see ◊nitrogen cycle).

crossing over process that occurs during ◊meiosis in which the paired chromosomes twist around each other and exchange corresponding chromosomal segments. It is a form of genetic ◊recombination, which increases variation and thus provides the raw material of evolution.

cuticle in zoology, the horny noncellular surface layer of many invertebrates such as insects; it reduces water loss and, in arthropods, acts as an ◊exoskeleton.

cuticle in botany, the waxy surface layer on those parts of the plant that are exposed to the air. It reduces water loss, and is continuous except for the stomata (see ◊stoma) and ◊lenticels.

cyanocobalamin chemical name for vitamin B_{12}, a ◊vitamin that is normally produced by microorganisms in the gut. The richest natural source is raw liver. The deficiency disease, pernicious ◊anaemia, results from the poor development of red blood cells. Sufferers develop extensive bruising and recover slowly from even minor injuries.

cytology the study of ◊cells and their functions. Major advances have been made possible in this field by the development of electron ◊microscopes.

cytoplasm the part of the cell outside the ◊nucleus. Strictly speaking, this includes all the ◊organelles (mitochondria, chloroplasts, and so on), but often cytoplasm refers to the jellylike matter in which the organelles are embedded (more correctly called the cytosol).

D

Darwin Charles Robert 1809–1882. English biologist who developed the modern theory of ◊evolution and proposed, with Alfred Russel Wallace, the principle of ◊natural selection.

As a young man, Darwin served as a naturalist on board the naval survey ship HMS *Beagle*. His travel experiences formed the basis of his theories, which were eventually pulished as *On the Origin of Species by Means of Natural Selection or the Preservation of Favoured Races in the Struggle for Life* 1859. His main argument, known as the theory of natural selection, concerned the variation existing between members of a sexually reproducing population. According to Darwin, those members with variations better fitted to the environment would be more likely to survive and breed, subsequently passing on these favourable characteristics to their offspring. Over time the population would change accordingly, and given enough time, a new species would form.

death the ending of all life functions, so that the molecules and structures associated with living things become disorganized and indistinguishable from similar molecules found in non-living things. Living organisms expend large amounts of energy preventing their complex molecules from breaking up; cellular repair and replacement are vital processes in multicellular organisms. At death this energy is no longer available, and the processes of disorganization become inevitable.

deciduous describing trees and shrubs that shed their leaves before the onset of winter or a dry season. In temperate regions there is little water available during winter, and leaf fall is an adaptation to reduce ◊transpiration. Examples of deciduous trees are oak and beech.

decomposer or ***saprotroph*** any organism that breaks down dead matter. Decomposers play a vital role in the ◊ecosystem by freeing important chemical substances, such as nitrogen compounds, locked up in dead organisms or excrement. They feed on some of the released organic matter, but leave the rest to filter back into the soil or pass in gas form into the atmosphere. The principal decomposers are bacteria and fungi, but earthworms and many other invertebrates are often included in this group. The ◊nitrogen cycle relies on the actions of decomposers.

decomposition the destruction of dead organisms either by chemical reduction or by the action of ◊decomposers.

deep freezing method of preserving food by rapid freezing and storage at –18°C. See ◊food technology.

deforestation the destruction of forest for timber or for agriculture, without planting new trees to replace those lost (reafforestation). Deforestation causes fertile soil to be blown away or washed into rivers, leading to ◊soil erosion, drought, and flooding.

denaturation irreversible changes occurring in the structure of proteins such as enzymes, usually caused by changes in pH or temperature. An example is the heating of egg albumen resulting in solid egg white.

The enzymes associated with digestion and metabolism become inactive if given abnormal conditions. Heat will damage their complex structure so that the usual interactions between enzyme and substrate can no longer occur.

dendrite slender filament projecting from the cell body of a ◊neuron (nerve cell) that receives incoming messages from other neurons and passes them on to the cell body. If the combined effect of these messages is strong enough, the cell body will send an electrical impulse along the ◊axon (filament that transmits messages), which will in turn pass its message to the dendrites of other neurons.

denitrification a process occurring naturally in soil by which bacteria break down ◊nitrates to give nitrogen gas, which returns to the atmosphere.

dental formula a way of showing what an animal's teeth are like. The dental formula consists of eight numbers separated by a line into two rows. The four above the line represent the teeth in one side of the upper jaw, starting at the front. If this reads 2 1 2 3 (as for humans) it means two incisors, one canine, two premolars, and three molars (see

◊tooth). The numbers below the line represent the lower jaw. The total number of teeth can be calculated by adding up all the numbers and multiplying by two.

dentition the type and number of teeth in a species. Different kinds of teeth have different functions, and a grass-eating animal will have well developed molars for grinding its food, whereas a meat-eater will need large canines for catching and killing its prey. The teeth that are less useful may be reduced in size or missing altogether. An animal's dentition is represented diagramatically by a ◊dental formula.

dentition

carnivore (dog)

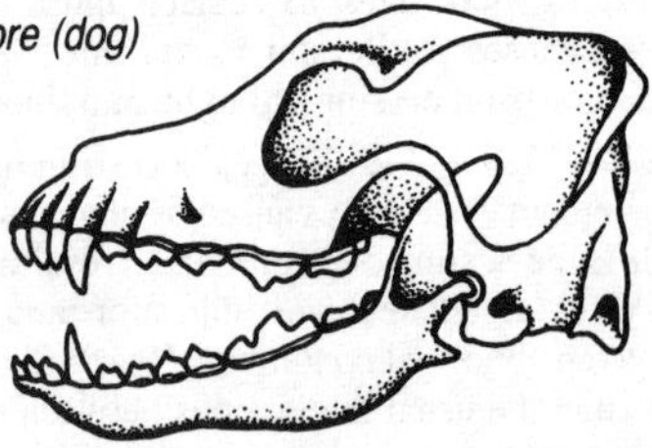

dental formula $i\frac{3}{3}\ c\frac{1}{1}\ pm\frac{4}{4}\ m\frac{2}{3}$

herbivore (sheep)

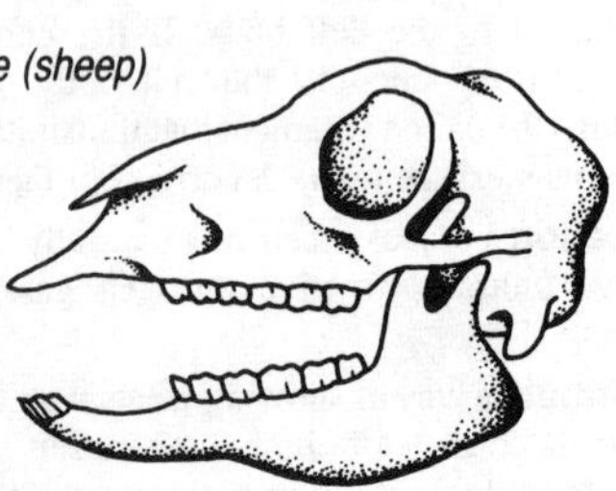

dental formula $i\frac{0}{3}\ c\frac{0}{0}\ pm\frac{3}{3}\ m\frac{3}{3}$

deoxyribonucleic acid the full name of ◊DNA.

destarching in photosynthesis experiments, the process by which all starch is removed from a plant so that the effects of increased light or carbon dioxide concentration can be studied. It is achieved by placing the plant in the dark for 24 hours. During this time the plant cannot photosynthesize and therefore uses up its reserves of food.

detritus the organic debris produced during the ◊decomposition of animals and plants.

development the process whereby a living thing transforms itself from a single cell into a vastly complicated multicellular organism, with structures such as limbs and functions such as respiration all able to work correctly in relation to each other. Most of the details of this process remain unknown, although some of the central features are becoming understood.

Apart from the sex cells (◊gametes), each cell within an organism contains exactly the same genetic code. Whether a cell develops into a liver cell or a brain cell depends therefore not on which genes it contains, but on which genes are allowed to be expressed. The development of forms and patterns within an organism, and the production of different, highly specialized cells, is a problem of control, with genes being turned on and off according to the stage of development reached by the organism. See ◊differentiation.

diabetes the disease *diabetes mellitus* in which a disorder of the ◊islets of Langerhans in the ◊pancreas prevents the body producing the hormone ◊insulin, so that sugars cannot be used properly. Treatment is by strict dietary control and oral or injected insulin. See ◊blood glucose regulation.

diaphragm in mammals, a muscular sheet separating the thorax from the abdomen. Its rhythmical movements affect the size of the thorax and cause the pressure changes within the lungs that result in ◊breathing.

diaphragm or ***cap*** a barrier ◊contraceptive that is pushed into the vagina and fits over the cervix (neck of the womb), preventing sperm from entering the womb (uterus). For a cap to be effective, a

◊spermicide must be used. This method is 97% effective if practised correctly.

dicotyledon a major subdivision of the angiosperms, containing the great majority of flowering plants. Dicotyledons are characterized by the presence of two seed leaves, or ◊cotyledons, in the embryo. They generally have broad leaves with netlike veins. Compare ◊monocotyledon.

diet the range of food eaten by an animal. The basic components of a diet are a group of chemicals: proteins, carbohydrates, fats, vitamins, minerals, and water. Different animals require these substances in different proportions, but the necessity of finding and processing an appropriate diet is a very basic drive in animal evolution.

The diet an animal needs may vary over its lifespan, according to whether it is growing, reproducing, highly active, or approaching death. For instance, an animal may need increased carbohydrate for additional energy, or increased minerals during periods of growth.

differentiation in developing embryos, the process by which cells become increasingly different and specialized, giving rise to more complex structures that have particular functions in the adult organism. For instance, embryonic cells may develop into nerve, muscle, or bone cells. See ◊development.

diffusion the random movement of molecules from a region in which they are at a high concentration to a region in which they are at a low concentration until an equal concentration is achieved throughout. The change in concentration that gives rise to diffusion is called a ***concentration gradient***. For instance, if sugar is added to water the sugar molecules will diffuse along the concentration gradient until they become evenly distributed throughout.

In biological systems, diffusion plays an essential role in the transport, over short distances, of molecules such as nutrients, respiratory gases (carbon dioxide and oxygen), and neurotransmitters. It provides the means by which small molecules pass into and out of individual cells and very small organisms, such as *Amoeba*, that possess no circulatory system. Compare ◊osmosis (a special type of diffusion) and ◊active transport. See also ◊gas exchange.

diffusion

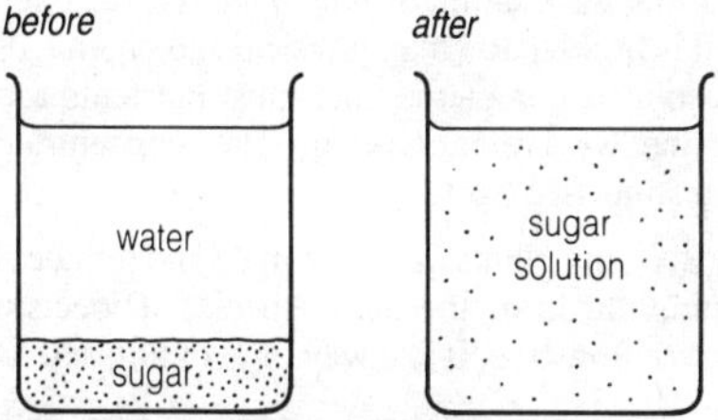

sugar and water molecules become evenly mixed

gas exchange in Amoeba

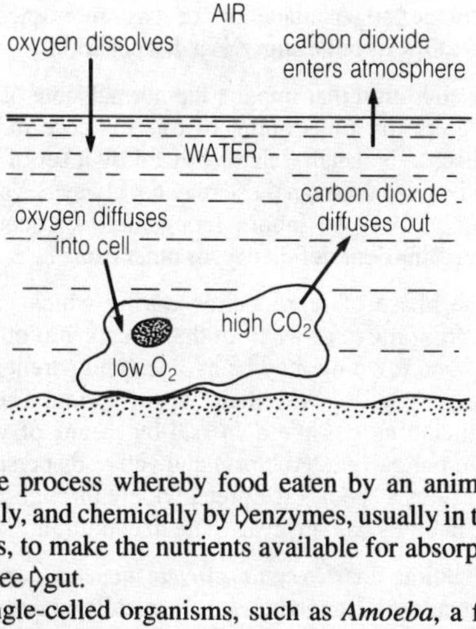

digestion the process whereby food eaten by an animal is broken down physically, and chemically by ◊enzymes, usually in the ◊stomach and ◊intestines, to make the nutrients available for absorption and cell metabolism. See ◊gut.

In some single-celled organisms, such as *Amoeba*, a food particle is engulfed by the cell itself, and digested in a ◊vacuole within the cell.

digestive system the mouth, stomach, small and large intestines, and associated glands of animals, which are responsible for digesting food. The food is broken down by physical and chemical means in the stomach; digestion is completed, and most nutrients are absorbed in the small intestine; what remains is stored and concentrated into faeces in the large intestine. See ◊gut.

dioecious describing plants that have male and female flowers borne on separate individuals of the same species. Dioecism occurs, for example, in the willow tree. It is a way of avoiding self-fertilization.

diploid having two sets of ◊chromosomes in each cell. In sexually reproducing species, one set is derived from each parent, the ◊gametes, or sex cells, of each parent being ◊haploid (having only one set of chromosomes) due to ◊meiosis (reduction cell division).

disaccharide a ◊sugar made up of two monosaccharide units. Sucrose $C_{12}H_{22}O_{11}$, or table sugar, is a disaccharide.

disease any condition that impairs the normal state of an organism, and usually alters the functioning of one or more of its organs or systems. A disease is usually characterized by a set of characteristic symptoms and signs, although these may not always be apparent to the sufferer. Diseases may be inborn (congenital) or acquired through infection, injury, nutrient deficiency, or other cause.

dispersal the phase of reproduction during which gametes, eggs, seeds, or offspring move away from the parents into other areas. The result is that overcrowding may be avoided and parents will not find themselves in competition with their own offspring. The mechanisms are various, including passive dispersal by means of wind or water currents or animal carriers (vectors) and active dispersal by locomotion. The ability of a species to spread widely through an area and to colonize new habitats has survival value in evolution.

DNA (abbreviation for ***deoxyribonucleic acid***) a complex double-stranded molecule that contains, in chemically coded form, all the information needed to build, control, and maintain a living organism. DNA is a ladderlike double-stranded ◊nucleic acid that forms the basis

DNA

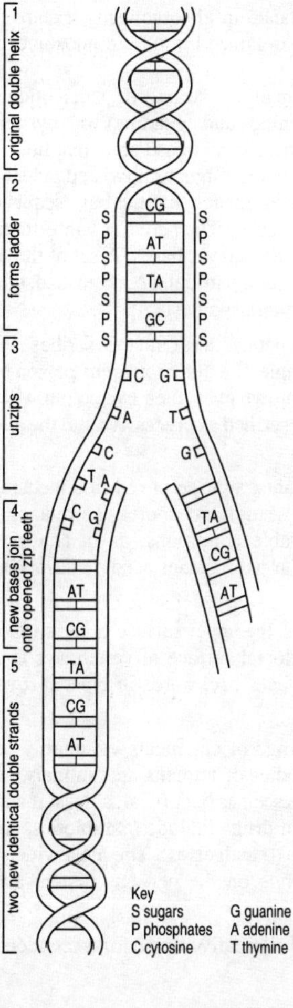
1 original double helix
2 forms ladder
3 unzips
4 new bases join onto opened zip teeth
5 two new identical double strands
S P S P S P S
S P S P S P S
CG
AT
TA
GC
C G
A T
C G
T A
C G
TA
CG
AT
AT
CG
TA
CG
AT
Key
S sugars
P phosphates
C cytosine
G guanine
A adenine
T thymine

of genetic inheritance in all organisms, except for a few viruses. In ◊eukaryotes, it is organized into ◊chromosomes and contained in the cell nucleus.

DNA is made up of two strands of ◊nucleotide sub-units which contain purine (adenine and guanine) or pyrimidine (cytosine and thymine) bases; these form ◊base pairs that link the two strands of the DNA molecule like the rungs of a twisted ladder. The specific way in which the pairs form means that the base sequence is preserved from generation to generation. The hereditary information is stored as a specific sequence of the various bases. A set of three bases is known as a codon and it specifies a particular ◊amino acid, the sub-unit of a protein molecule; see ◊protein synthesis.

dominant in genetics, term that describes an ◊allele that masks another. For example, if a ◊heterozygous person has one allele for blue eyes and one for brown eyes, their eye colour will be brown. The allele for blue eyes is described as ◊recessive and the allele for brown eyes as dominant.

dormancy in plants, a phase of reduced metabolic activity exhibited by certain buds, seeds, and spores. Dormancy can help a plant to survive unfavourable conditions, as in annual plants that pass the cold winter season as dormant seeds, and plants that form dormant buds.

dorsal in animals, the upper surface or the surface furthest away from the ground. The dorsal surface of vertebrates is the surface closest to the backbone; it faces backwards in bipedal (two-legged) vertebrates such as humans.

drug any of a range of chemicals voluntarily or involuntarily introduced into the bodies of humans and animals. Most drugs in use are medicines and are seen as beneficial, at least if administered in the correct amount. Such drugs include ◊antibiotics, stimulants, ◊sedatives, and pain-relievers (◊analgesics). The most widely-used drugs, tobacco and alcohol, operate on the nervous system and can be considered dangerous.

ductless gland alternative name for an ◊endocrine gland.

duodenum short length of gut (alimentary canal) found between the stomach and the ileum. Its role is in the digestion of carbohydrates, fats, and proteins. The smaller molecules formed are then absorbed, either in the duodenum itself or in the ileum.

Entry to the duodenum is controlled by the ***pyloric sphincter***, a muscular ring at the base of the stomach. Once food has passed into the duodenum it is mixed with bile from the liver and with a range of enzymes secreted from the pancreas, a digestive gland near the top of the intestine. The bile neutralizes the acidity of the gastric juices passing out of the stomach and aids fat digestion.

E

ear the organ of hearing in animals. It responds to the vibrations that constitute sound, and these are translated into nerve signals and passed to the brain. A mammal's ear consists of three parts: outer ear, middle ear, and inner ear. The ***outer ear*** is a funnel that collects sound, directing it down a tube to the ***ear drum*** (tympanic membrane), which separates the outer and ***middle ear***. Sounds vibrate this membrane, the mechanical movement of which is transferred to a smaller membrane leading to the ***inner ear*** by three small bones, the auditory ossicles. Vibrations of the inner ear membrane move fluid contained in the snail-shaped cochlea, which vibrates hair cells that stimulate the auditory nerve, connected to the brain. Three fluid-filled canals of the inner ear detect changes of position; this mechanism, with other sensory inputs, is responsible for the sense of balance.

The ear is usually on the head, but in some insects is found on the legs, thorax, or abdomen.

ecology (Greek *oikos* 'house') the study of the relationship between an organism and the environment in which it lives, including other living organisms and the nonliving surroundings.

Ecology may be concerned with individual organisms, with populations or species, or with entire ◊communities (for example, competition between species for access to resources in an ecosystem, or predator–prey relationships). Applied ecology is concerned with the management and conservation of habitats and the consequences and control of pollution.

ecosystem in ecology, a unit made up of a group (community) of living things interacting with the physical environment It is therefore made up of two components: the ***biotic***, or living, and the ***abiotic***, or non-living. The relationship between all these components is finely balanced, and the continuation of that balance is essential for the main-

ear

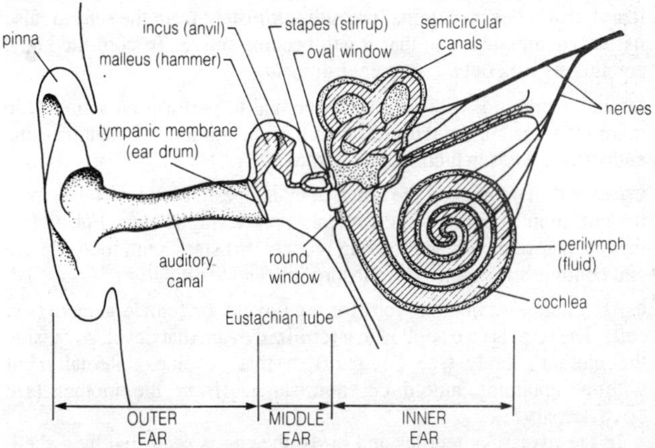

tenance of the ecosystem. The alteration of any one component can have a disastrous effect—for instance, the removal of a major predator might allow the number of herbivores to rise to such a level that the vegetation is severely damaged.

Energy and nutrients from the abiotic component of the ecosystem pass through the organisms of the biotic component in a particular sequence (see ◊food chain). The Sun's energy is captured through ◊photosynthesis, and nutrients are taken up from the soil or water by green plants (primary producers); both are passed to the herbivores that eat the plants and then to carnivores that feed on herbivores. These nutrients are returned to the soil through the ◊decomposition of excrement and dead organisms, thus completing a cycle that is crucial to the ecosystem's stabililty and survival.

Ecosystems may be aquatic (as in lakes and rockpools) or terrestrial (as in forests and grasslands).

ectoparasite ◊parasite that lives on the outer surface of its host.

ectoplasm outer layer of a cell's ◊cytoplasm.

ectotherm a 'cold-blooded' animal (see ◊poikilothermy), such as a lizard, that relies on external warmth (ultimately from the sun) to raise its body temperature so that it can become active. To cool the body, ectotherms seek out a cooler environment.

effector organ or cell used by an animal to perform an action. The main effectors are the muscles, which bring about movement, and the endocrine glands, which secrete hormones.

egestion or ***defecation*** the removal of undigested food or faeces from the gut. In most animals egestion takes place via the anus. Egestion is the last part of a complex feeding process that starts with food capture and continues with digestion, absorption, and assimilation.

egg in animals, another word for the ◊ovum, or female gamete (sex cell). The term is also applied to a fertilized ovum that develops outside the mother's body (see ◊ovipary) or that develops internally but without obtaining any direct nourishment from the mother (see ◊ovovivipary).

In the oviparous reptiles and birds, the egg is protected by a shell, and well supplied with nutrients in the form of yolk.

egg cell in plants, another word for the ◊ovum or oosphere, the female gamete.

embryo early development stage of an animal or a plant following fertilization of an ovum (female gamete), or activation of an ovum by ◊parthenogenesis.

In animals the embryo exists either within an egg (where it is nourished by food contained in the yolk), or, in mammals, in the ◊uterus of the mother. In mammals (except marsupials) the embryo is fed through the ◊placenta. In humans the term embryo describes the fertilized egg during its first seven weeks of existence; from the eighth week onwards it is referred to as a fetus. The plant embryo is found within the seed in higher plants. It sometimes consists of only a few cells, but usually includes a root; a shoot (or primary bud); and one or two ◊cotyledons, which nourish the growing seedling.

emulsifaction the process in the small intestine by which ◊bile, secreted from the liver, breaks down large globules of fat into microscopically small particles. These tiny droplets, which are less than 0.5 micrometres in diameter, can then be attacked by the digestive enzyme lipase.

emulsifier a food ◊additive used to keep oils dispersed and in suspension, in products such as mayonnaise and peanut butter. Egg yolk is a naturally occurring emulsifier.

endangered species plant or animal species whose numbers are so few that it is at risk of becoming extinct.

An example of an endangered species is the Javan rhinoceros. There are only about 50 alive today and, unless active steps are taken to promote this species' survival, it will probably be extinct within a few decades.

endocrine gland gland that secretes ◊hormones into the bloodstream in order to regulate body processes. In humans the main endocrine glands are the pituitary, thyroid, adrenal, pancreas, ovary, and testis. Compare ◊exocrine gland.

endolymph fluid found in the inner ◊ear; it fills the central passage of the cochlea as well as the semicircular canals.

Sound waves travelling into the ear pass eventually through the three small bones of the middle ear and set up vibrations in the endolymph. These are detected by receptors in the cochlea, which send nerve impulses to the hearing centres of the brain.

endoparasite ◊parasite that lives inside the body of its host.

endoplasm inner, liquid part of a cell's ◊cytoplasm.

endoplasmic reticulum (ER) a membranous network in eukaryotic cells. It stores and transports proteins needed elsewhere in the cells and also carries various enzymes needed for the manufacture of ◊fats. The ◊ribosomes, which carry out ◊protein synthesis, are attached to parts of the ER.

endoskeleton the internal supporting structure of vertebrates, made up of cartilage or bone. It provides support, and acts as a system of levers to which muscles are attached to provide movement. Certain parts of the skeleton (the skull and ribs) give protection to vital body organs.

endocrine gland

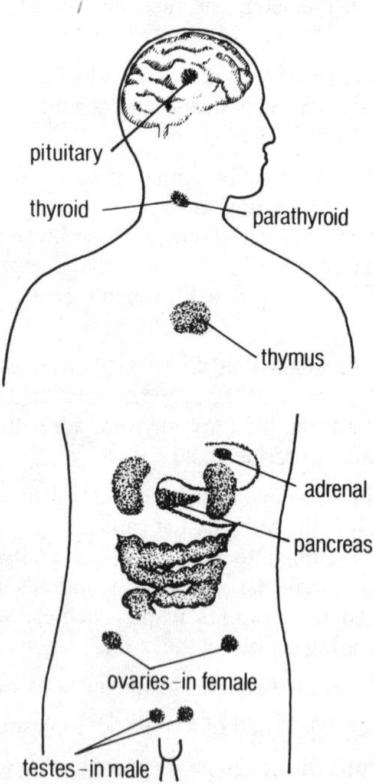

endosperm a nutritive tissue in the seeds of most flowering plants. It surrounds the embryo and is produced by an unusual process that parallels the ◊fertilization of the ovum by a male gamete. A second male gamete from the pollen grain fuses with two female nuclei within the ◊embryo sac. Thus endosperm cells are triploid (having three sets of chromosomes); they contain food reserves such as starch, fat, and protein that are utilized by the developing seedling.

endotherm a 'warm-blooded', or homeothermic animal. Endotherms have internal mechanisms for regulating their body temperatures to levels different from the environmental temperature. See ◊homeothermy.

environment in ecology, the sum of conditions affecting a particular organism or ◊ecosystem, including physical surroundings, climate, and influences of other living organisms. See also ◊biosphere and ◊habitat.

environmentalism a theory emphasising the primary influence of the environment on the development of groups or individuals. It stresses the importance of the physical, biological, psychological or cultural environment as a factor influencing the structure or behaviour of animals, including humans.

enzyme a biological ◊catalyst produced in cells, and capable of speeding up the chemical reactions necessary for life by converting one molecule (substrate) into another. Enzymes are not themselves destroyed by this process. They are large, complex ◊proteins, and are highly specific, each particular substrate requiring its own particular enzyme. The enzyme fits into a 'slot' (active site) in the substrate molecule forming an enzyme–substrate complex that lasts until the substrate is altered or split, after which the enzyme can fall away. The substrate may therefore be compared to a lock, and the enzyme to the key required in order to open it.

The lock-and-key model is an attempt to explain the activity of enzymes. The substrate is seen as a padlock; the enzyme is the key. Only one type of key is the right shape to fit the lock. Once the padlock has been split, the key can be removed unchanged and can repeat the process.

The activity and efficiency of enzymes are influenced by various factors, including temperature and pH conditions. Temperatures above 60°C damage (denature) the intricate structure of enzymes, causing reactions to cease. Each enzyme operates best within a specific pH range, and is denatured by excessive acidity or alkalinity.

Digestive enzymes include ◊amylases (which digest starch), ◊lipases (which digest fats), and ◊proteases (which digest protein). Other

enzyme

lock-and-key model of enzyme activity

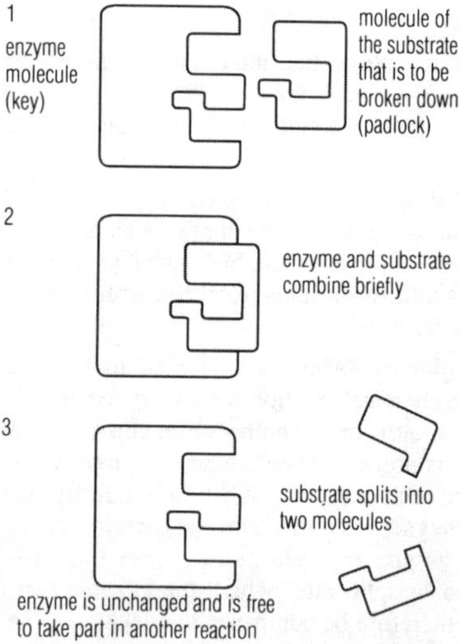

enzymes play a part in the conversion of food energy into ◊ATP; the manufacture of all the molecular components of the body; the replication of ◊DNA when a cell divides; the production of hormones; and the control of movement of substances into and out of cells.

Enzymes have many medical and industrial uses, from washing powders to drug production, and as research tools in molecular biology. They can be extracted from bacteria and moulds, and ◊genetic engineering now makes it possible to tailor an enzyme for a specific purpose.

epidermis the outermost layer of cells on an organism's body. In plants and many invertebrates such as insects, it consists of a single layer of cells. In vertebrates, it consists of several layers of cells.

The epidermis of plants and invertebrates often has an outer noncellular ◊cuticle that protects the organism from dehydration. In vertebrates such as reptiles, birds, and mammals, the outermost layer of cells is dead, forming a tough, waterproof layer known as ◊skin.

epiglottis a small cartilaginous flap in the pharynx; it moves during swallowing to prevent food from passing into the trachea (windpipe) and causing choking.

The action of the epiglottis is a highly complex reflex process involving two phases. During the first stage a mouthful of chewed food is lifted by the tongue towards the top and back of the mouth. This is accompanied by the stopping of breathing and by the blocking of the nasal areas from the mouth. The second phase involves the epiglottis moving over the larynx while the food passes down into the oesophagus.

epithelium in animals, closely packed cells that form a surface. It may be protective, as in skin, or secretory, as in the wall of the gut.

The disease scurvy, caused by a lack of vitamin C, affects the health of epithelial surfaces including the gums.

erythrocyte another name for ◊red blood cell.

ethanol C_2H_5OH alcohol found in beer, wine, cider, spirits, and other alcoholic drinks. It is produced naturally by the fermentation of carbohydrates by yeast cells (see ◊anaerobic respiration).

etiolation form of growth seen in plants receiving insufficient light. It is characterized by long, weak stems, small leaves, and a pale yellowish colour (◊chlorosis) owing to a lack of chlorophyll. The rapid increase in height enables a plant that is surrounded by others to quickly reach a source of light, after which return to normal growth usually occurs.

eukaryote one of the two major groups into which organisms are divided. The eukaryotes comprise all living things, except bacteria and cyanobacteria, which belong to the ◊prokaryote group.

The cells of eukaryotes, unlike those of prokaryotes, possess a clearly defined nucleus, bounded by a membrane, within which DNA is formed into distinct chromosomes. Eukaryotic cells also contain mitochondria, chloroplasts, and other structures (organelles).

Eustachian tube small air-filled canal connecting the middle ◊ear with the back of the throat. It is found in all land vertebrates, and serves to equalize the pressure on both sides of the tympanic membrane (ear drum).

eutrophication the excessive enrichment of lake waters, primarily by artificial, nitrate fertilizers, washed from the soil by rain, and by phosphates from detergents. These encourage the growth of algae and bacteria, which use up the oxygen in the water making it uninhabitable for fishes and other animal life.

evergreen plant, such as pine, spruce, or holly, that bears its leaves all year round. Most conifers are evergreen. Plants that shed their leaves in autumn or a dry season are described as ◊deciduous.

evolution a slow process of change from one form to another, as in the evolution of the universe from its formation in the Big Bang to its present state, or in the evolution of life on Earth. Some Christians and Muslims deny the theory of evolution as conflicting with the belief that God created all things (◊creationism).

With respect to the living world, the idea of continuous evolution can be traced as far back as the 1st century BC, but it did not gain wide acceptance until the 19th century following the work of Charles ◊Darwin. Darwin assigned the major role in evolutionary change to ◊natural selection acting on randomly occurring variations (now known to be produced by spontaneous changes or ◊mutations in the genetic material of organisms). Natural selection occurs because those individuals better adapted to their particular environments reproduce more effectively, thus contributing their characteristics (in the form of genes) to future generations. The current theory of evolution, called ◊Neo-Darwinism, combines Darwin's theory with Gregor Mendel's theories on genetics (see ◊Mendelism). See also ◊sexual selection and ◊adaptive radiation.

excretion the removal of waste products from the cells of living organisms. In plants and simple animals, waste products are removed by diffusion, but in higher animals specialized organs are required. In mammals, for example, carbon dioxide and water are removed via the lungs, and nitrogenous compounds and water via the liver, the kidneys, and the rest of the urinary system.

exocrine gland gland that discharges secretions, usually through a tube or a duct, onto a surface. Examples include sweat glands which release sweat onto the skin, and digestive glands which release digestive juices onto the walls of the intestine. Compare ◊endocrine gland.

exoskeleton the hardened external skeleton of insects, spiders, crabs, and other arthropods. It provides attachment for muscles and protection for the internal organs, as well as support. To permit growth it is periodically shed in a process called ecdysis.

extensor a muscle that straightens a limb.

extinction the complete disappearance of a species. In the past, extinctions are believed to have occurred because species were unable to adapt quickly enough to a naturally changing environment. Today, most extinctions are due to human activity. Some species, such as the dodo of Mauritius, the moas of New Zealand, and the passenger pigeon of North America, were exterminated by hunting. Others become extinct when their habitat is destroyed. See also ◊endangered species.

Mass extinctions are episodes during which whole groups of species become extinct, the best known being that of the dinosaurs, other large reptiles, and various marine invertebrates about 65 million years ago.

extracellular spaces in an organism that are not cellular. Such areas include the interstitial fluid found between cells, the hollow region of the gut, and the canal passing down the centre of the spinal cord.

eye the organ of vision. The human eye is a roughly spherical structure contained in a bony socket. Light enters it through the ***cornea***, and passes through the circular opening (***pupil***) in the iris (the coloured part of the eye). The light is focused by the combined action of the curved cornea, the internal fluids (aqueous and vitreous humours), and the ***lens*** (the rounded transparent structure behind the iris). The ciliary muscles act on the lens to change its shape, so that images of objects at different distances can be focused on the ◊***retina***. This is at the back of the eye, and is packed with light-sensitive cells (rods and cones), connected to the brain by the optic nerve.

The ***compound eye*** found in insects is made of many separate facets, each of which collects light and directs it separately to a receptor to build up an image. The ***simple eye*** (ocellus) found in invertebrates such as worms and snails consists only of a small group of light-sensitive cells and a lens made of cuticle. Squids and octopuses have complex eyes similar to those of vertebrates.

F

faeces remains of food and other debris passed out of the digestive tract of animals. Faeces consist of quantities of cellulose material, bacteria and other microorganisms, rubbed-off lining of the digestive tract, bile fluids, undigested food, minerals, and water.

Fallopian tube or ***oviduct*** in mammals, one of two tubes that carry ova (eggs) from the ovary to the uterus. An ovum is fertilized by sperm in the Fallopian tubes, which are lined with cells whose ◊cilia move the ovum towards the uterus.

family in classification, a group of related genera (see ◊genus). Family names are not printed in italic (unlike genus and species names). Related families are grouped together in an ◊order.

family planning spacing or preventing the birth of children. Access to family-planning services (see ◊contraceptive) is a significant factor in women's health as well as in limiting population growth. If all those women who wished to avoid further childbirth were able to do so, the number of births would be reduced by 27% in Africa, 33% in Asia, and 35% in Latin America; and the number of women who die during pregnancy or childbirth would be reduced by about 50%.

The average number of pregnancies per woman is two in the industrialized countries, where 71% use family planning, as compared to six or seven pregnancies per woman in the Third World. According to a World Bank estimate, doubling the annual $2 billion spent on family planning would prevent the deaths of 5.6 million infants and 250,000 mothers each year.

fat in the broadest sense, a mixture of ◊lipids—chiefly triglycerides (lipids containing three ◊fatty acid molecules linked to a molecule of glycerol). More specifically, the term refers to a lipid mixture that is solid at room temperature (20°C); lipid mixtures that are liquid at room temperature are called ***oils***. The higher the proportion of saturated fatty acids in a mixture the harder the fat.

Fats are essential sources of energy for many animals; they have a ◊calorific value twice that of carbohydrates, but are more difficult to digest and respire. In many animals and plants, excess carbohydrates and proteins are converted into fats for storage. Vertebrates, particularly mammals, store fats in specialized connective tissues (◊adipose tissues), which not only act as energy reserves but also insulate the body and cushion body organs.

fatty acid organic acid consisting of a hydrocarbon chain, up to 24 carbon atoms long, with a carboxyl group (–COOH) at one end.

The bonds may be single or double; where a double bond occurs the carbon atoms concerned carry one instead of two hydrogen atoms. Chains with only single bonds have all the hydrogen they can carry, so they are said to to be ***saturated*** with hydrogen. Chains with one or more double bonds are said to be ***unsaturated*** (see ◊polyunsaturates). Fatty acids are generally found combined with glycerol in tryglycerides (see ◊fat).

feather a rigid outgrowth of the outer layer of the skin of birds, made of the protein keratin. Feathers provide insulation and facilitate flight. There are several types, including long quill feathers on the wings and tail, fluffy down feathers for retaining body heat, and contour feathers covering the body. The colouring of feathers is often important in camouflage or in courtship and other displays. Feathers are replaced at least once a year.

fecundity the rate at which an organism reproduces, as distinct from its ability to reproduce (◊fertility). In vertebrates, it is usually measured as the number of offspring produced by a female each year.

feedback another term for ◊biofeedback.

femur or ***thigh-bone*** the upper bone in the hind limb of a four-limbed vertebrate.

fermentation the breakdown of sugars by bacteria and yeasts using a method of respiration without oxygen (◊anaerobic respiration). Fermentation processes have long been used in baking bread, making beer and wine, and producing cheese, yoghurt, soy sauce, and many other foodstuffs.

In baking and brewing, yeasts ferment sugars to produce ethanol (alcohol) and carbon dioxide; the latter makes bread rise and puts bubbles into beers and champagne.

fertility an organism's ability to reproduce, as distinct from the rate at which it reproduces (see ◊fecundity). Individuals become infertile (unable to reproduce) when they cannot generate gametes (eggs or sperm) or when their gametes cannot yield a viable ◊embryo after fertilization.

fertilization or ***conception*** in sexual reproduction, the union of two ◊gametes (sex cells, often called ovum and sperm) to produce a ◊zygote, which combines the genetic material contributed by each parent. In ***self-fertilization*** the male and female gametes come from the same plant; in ***cross-fertilization*** they come from different plants. Self-fertilization rarely occurs in animals; usually even ◊hermaphrodite animals cross-fertilize each other.

fertilization

fertilization of a flowering plant

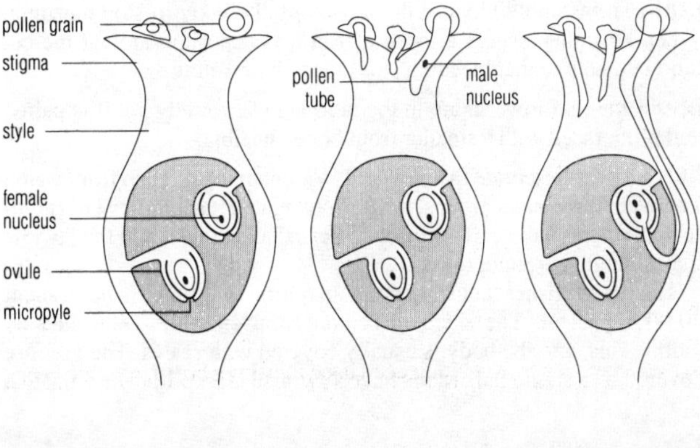

In insects, mammals, reptiles and birds, fertilization occurs within the female's body; in the majority of fish and amphibians, and most aquatic invertebrates, it occurs externally, when both sexes release their gametes freely into the water. In most fungi, gametes are not released, but the hyphae of the two parents grow towards each other and fuse to achieve fertilization. In higher plants, ◊pollination precedes fertilization.

fertilizer substance containing a range of about 20 chemical elements necessary for healthy plant growth, used in agriculture and horticulture to compensate for the deficiencies of poor or depleted soil, and improve plant growth. Fertilizers may be ***organic*** (derived from living things) – for example, manure, compost, or ashes; or ***inorganic*** (artificial), mainly in the form of compounds of nitrogen, potassium, and phosphate, which have been used on a very much increased scale since 1945.

Artificial fertilizers are often in excess of plant requirements and tend to leach away to pollute lakes and rivers; see ◊eutrophication.

fetus or ***foetus*** a stage in mammalian ◊embryo development. The human embryo is usually called a fetus after the eighth week of development, when the limbs and external features of the head are recognizable.

fibrin an insoluble blood protein used by the body to stop bleeding. When an injury occurs a mesh of fibrin is deposited around the cut blood vessels so that bleeding stops. See ◊blood clotting.

fibula the rear lower bone in the hind leg of a vertebrate. It is paired and often fused with a smaller front bone, the tibia.

fish aquatic vertebrate that uses gills for obtaining oxygen from water. There are three main groups, not closely related: the bony fishes (goldfish, cod, tuna), the cartilaginous fishes (sharks, rays), and the jawless fishes (hagfishes, lampreys).

The bony fishes constitute the majority of living fishes (about 20,000 species). The skeleton is bone, movement is controlled by mobile fins, and the body is usually covered with scales. The gills are covered by a single flap. Many have a swim bladder with which the fish adjusts its buoyancy. Most lay eggs, sometimes in vast numbers; some cod can produce as many as 28 million.

fitness in genetic theory, a measure of the success with which a genetically determined character can spread in future generations. By convention, the normal character is assigned a fitness of one, and variants (determined by other ◊alleles) are then assigned fitness values relative to this. Those with fitness greater than one will spread more rapidly and will ultimately replace the normal allele; those with fitness less than one will gradually die out. See ◊natural selection.

flaccidity the loss of rigidity (turgor) in plant cells, caused by loss of water from the central vacuole so that the cytoplasm no longer pushes against the cellulose cell wall. If this condition occurs throughout the plant then wilting is seen.

Flaccidity can be brought about in the laboratory by immersing the plant cell in a strong salt or sugar solution. Water leaves the cell by ◊osmosis causing the vacuole to shrink. In extreme cases the actual cytoplasm pulls away from the cell wall, a phenomenon known as ***plasmolysis***.

flagellum a small hairlike organ on the surface of certain cells. Flagella are the motile organs of certain protozoa and single-celled algae, and of the sperm cells of higher animals. Unlike ◊cilia, flagella usually occur singly or in pairs; they are also longer and have a more complex whiplike action.

flexor any muscle that bends a limb. Flexors usually work in opposition to other muscles, the extensors, an arrangement known as antagonistic pairing.

flocculation in soils, the artificially-induced coupling together of particles to improve aeration and drainage. Clay soils, which have very tiny particles and are difficult to work, are often treated in this way. The method involves adding lime to the soil.

floral diagram a diagram showing the arrangement and number of parts in a flower, drawn in cross-section. An ovary is drawn in the centre, surrounded by representations of the other floral parts, indicating the position of each at its base. If any parts such as the petals or sepals are fused, this is also indicated. Floral diagrams allow the structure of different flowers to be compared, and are usually shown with the ◊floral formula.

floral formula a symbolic representation of the structure of a flower. Each kind of floral part is represented by a letter (K for calyx, C for corolla, P for perianth, A for androecium, G for gynoecium) and a number to indicate the quantity, for example, C5 for a flower with five petals. The number is in brackets if the parts are fused.

flower the reproductive unit of an angiosperm (flowering plant), typically consisting of four whorls of modified leaves: ◊sepals, ◊petals, ◊stamens, and ◊carpels. These are borne on a central axis or ◊receptacle. The many variations in size, colour, number and arrangement of parts are closely related to the method of pollination. Flowers adapted for wind pollination typically have reduced or absent petals and sepals and long, feathery ◊stigmas that hang outside the flower to trap airborne pollen. In contrast, the petals of insect-pollinated flowers are usually conspicuous and brightly coloured.

The sepals and petals are collectively known as the calyx and corolla respectively and together comprise the perianth with the function

flower

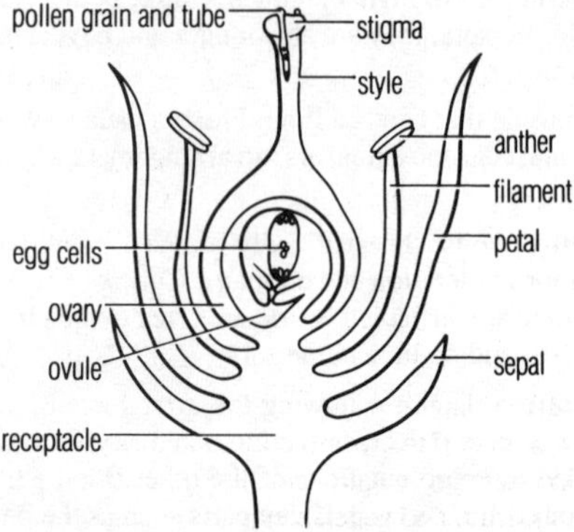

of protecting the reproductive organs and attracting pollinators. The stamens lie within the corolla, each having a slender stalk, or filament, bearing the pollen-containing anther at the top. Collectively they are known as the androecium. The inner whorl of the flower comprises the carpels, each usually consisting of an ◊ovary in which are borne the ◊ovules, and a stigma borne at the top of a slender stalk, or style. Collectively the carpels are known as the gynoecium. The stalk of a flower is called a pedicel.

Flowers may either be borne singly or grouped together in inflorescences. A flower is called hermaphrodite when it contains both male and female reproductive organs. When male and female organs are carried in separate flowers, they are termed monoecious; when male and female flowers are on separate plants, the term dioecious is used.

flowering plant common name for ◊angiosperm, a plant bearing flowers with various parts, including sepals, petals, stamens and carpels.

fluoridation addition of small amounts of fluoride salts to drinking water by certain water authorities to help prevent tooth decay. In areas where fluoride ions are naturally present in the water, research found that the incidence of tooth decay in children from those areas was reduced by more than 50%. A concentration of one part per million is sufficient to produce this beneficial effect.

follicle small group of cells that surround and nourish a structure such as a hair (hair follicle) or a cell such as an egg (Graafian follicle; see ◊menstrual cycle).

follicle-stimulating hormone (FSH) a hormone produced by the pituitary gland. It affects the ovaries in women, triggering off the production of an ovum (egg). Luteinizing hormone is needed to complete the process. In men, FSH stimulates the testes to produce sperm.

food anything eaten by human beings and other animals to sustain life and health. The building blocks of food are called nutrients, and humans can utilize the following nutrients:

carbohydrate as starch found in bread, potatoes, and pasta; as simple sugars in sucrose (table sugar) and honey; and as cellulose (dietary fibre) in cereals, fruit, and vegetables;

protein, of which good sources are nuts, fish, meat, eggs, milk, and some vegetables;

fat, found in most meat products, fish, dairy products (butter, milk, cheese), margarine, nuts, vegetable oils, and lard;

vitamins, found in a wide variety of foods, except for vitamin B_{12} which is only found in animal products;

minerals, found in a wide variety of foods; a good source of calcium is milk, of iodine is seafood, and of iron is liver or green vegetables;

water, found everywhere in nature;

alcohol, found in alcoholic beverages, from 40% in spirits to 0.01% in low-alcohol lagers and beers.

Food is needed for both its calorific energy, measured in kilojoules, and nutrients, such as proteins, that are converted to body tissues. Some nutrients mainly provide energy, such as fat, carbohydrate, and alcohol; other nutrients are important in other ways, such as aiding metabolism.

food capture the method by which an animal obtains food. There is no typical method, because food may exist as liquid, as tiny particles, or as large lumps. All animals show adaptations to their method of food capture. Mosquitoes have long sucking mouthparts, oysters have special flaps that swirl particles towards their mouth, and tigers have special ◊dentition for killing prey.

food chain in ecology, a way of showing feeding relations between plants and animals. Energy in the form of food is shown to be transferred from producer (the plants) to a series of consumers (first a herbivore, then a carnivore). In reality, organisms have varied diets, relying on different kinds of foods, so that the food chain is an over-simplification. The more complex ***food web*** shows a greater variety of relationships, but again emphasizes that energy passes from plants to herbivores to carnivores.

Environmental groups have used the concept of the food chain to show how poisons and other forms of pollution can pass from one animal to another.

food irradiation a development in ◊food technology, whereby food

food chain

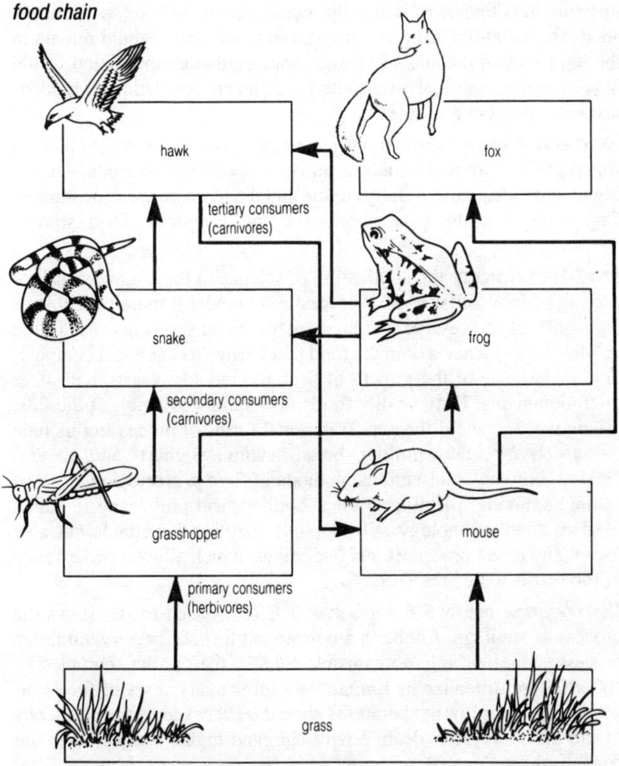

is exposed to low-level radiation to kill microorganisms.

Irradiation is highly effective, and does not make the food any more radioactive than it is naturally. Some vitamins are partially destroyed, such as vitamin C, and it would be unwise to eat only irradiated fruit and vegetables. The main cause for concern is that it may be used by unscrupulous traders to 'clean up' consignments of food, particularly shellfish, with high bacterial counts. Bacterial toxins would remain in the food, so that it could still cause illness, although irradiation would have removed signs of live bacteria. Stringent regulations would be needed to prevent this happening.

food store place in which nutrients can be stored; examples include starch grains within chloroplasts and endosperm within seeds. Among higher animals, adipose (fatty) tissue and the liver are used for storage. Organisms tend to have short-term and long-term food storage measures.

food technology the application of science to the commercial processing of foodstuffs. Food is processed to render it more palatable or digestible, or to preserve it from spoilage. Food spoils because of the actions of ◊enzymes within the food that change its chemical composition, or because of the growth of bacteria, moulds, yeasts, and other microorganisms. Fatty or oily foods also suffer oxidation of the fats, giving them a rancid flavour. Traditional forms of processing include boiling, frying, flour-milling, bread-making, yoghurt- and cheese-making, brewing, and various methods of ***food preservation***, such as salting, smoking, pickling, drying, bottling, and preserving in sugar. Modern food technology still employs traditional methods but also uses many novel processes and ◊additives, which allow a wider range of foodstuffs to be preserved.

Refrigeration below 5°C (or below 3°C for cooked foods) slows the process of spoilage. Although a convenient form of preservation, this process cannot kill microorganisms, nor stop their growth completely, and a failure to realize its limitations causes many cases of food poisoning. Refrigerator temperatures should be checked as the efficiency of the machinery can decline with age, and higher temperatures are dangerous.

Deep freezing (–18°C or below) stops almost all spoilage processes, although there may be some residual enzyme activity in uncooked vegetables, which is why these are blanched (dipped in hot water to destroy the enzymes) before freezing. Microorganisms cannot grow or divide, but most remain alive and can resume activity once defrosted.

Pasteurization is used mainly for milk. By holding the milk at a high temperature, but below boiling point, for a period of time, all disease-causing bacteria can be destroyed. The milk is held at 72°C for 15 seconds. Other, less harmful bacteria survive, so the milk will still go sour within a few days. Boiling the milk would destroy all bacteria, but impair the flavour.

Ultra-heat treatment is used to produce UHT milk. This process uses higher temperatures than pasteurization, and kills all bacteria present, giving the milk a long shelf life but altering the flavour.

Drying is an effective method of preservation because both microorganisms and enzymes need water to be active. Products such as dried milk and instant coffee are made by spraying the liquid into a rising column of dry, heated air.

Freeze-drying is carried out in a vacuum. It is less damaging to food than straight dehydration, and is used for quality instant coffee and dried vegetables.

Canning relies on high temperatures to destroy microorganisms and enzymes. The food is sealed into a can to prevent any recontamination by bacteria. Beverages may also be canned to preserve the carbon dioxide that makes drinks fizzy.

Pickling makes use of acetic acid, found in vinegar, to stop the growth of moulds.

Irradiation is a method of preserving food by subjecting it to low-level radiation. It is highly controversial (see ◊food irradiation) and not yet widely used in the UK.

Chemical treatments are widely used, for example in margarine manufacture, where hydrogen is bubbled through vegetable oils in the presence of a ◊catalyst to produce a more solid, spreadable fat. The catalyst is later removed. Chemicals that are introduced in processing and remain in the food are known as ***food additives*** and include flavourings, preservatives, antioxidants, emulsifiers, and colourings.

food test any of several types of simple test, easily performed in the laboratory, used to identify the main classes of food.

starch–iodine test Food is ground up in distilled water and iodine is added. A dense black colour indicates that starch is present.

sugar–Benedict's test Food is ground up in distilled water and placed in a test tube with Benedict's reagent. The tube is then heated in a boiling water bath. If a ◊reducing sugar, such as glucose, is present the colour changes from blue to brick-red; if a non-reducing sugar, such as sucrose, is present, there is no colour change.

protein–Biuret test Food is ground up in distilled water and a mixture of copper(II) sulphate and sodium hydroxide is added. If protein is present a mauve colour is seen.

fossil fuel fuel, such as coal or oil, formed from the fossilized remains of plants that lived hundreds of millions of years ago. Fossil fuels are a non-renewable resource and will run out eventually. Extraction of coal causes considerable environmental pollution, and burning coal contributes to problems of ◊acid rain and the ◊greenhouse effect.

freeze-drying method of preserving food; see ◊food technology. The product to be dried is frozen and then put in a vacuum chamber that forces out the ice as water vapour, a process known as sublimation.

Many of the substances that give products such as coffee their typical flavour are volatile, and would be lost in a normal drying process because they would evaporate along with the water. In the freeze-drying process these volatile compounds do not pass into the ice that is to be sublimed, and are therefore largely retained.

fructose ($C_6H_{12}0_6$) monosaccharide sugar that occurs naturally in honey, the nectar of flowers, and many sweet fruits.

fruit structure that develops from the carpel of a flower and encloses one or more seeds. Its function is to protect the seeds during their development and to aid in their dispersal. Fruits are often sweet, juicy, and colourful in order to attract animals to eat them and thus disperese the seeds.

The fruit structure consists of the ◊pericarp or fruit wall, usually divided into a number of distinct layers. Sometimes parts other than the ovary are incorporated into the fruit structure, resulting in a false fruit or ◊pseudocarp, such as the apple and strawberry. Fruits may open to shed their seeds (dehiscent) or remain unopened and be dispersed as a single unit (indehiscent).

Simple fruits (for example, peaches) are derived from a single ovary, whereas composite or multiple fruits (for example, blackberries) are formed from the ovaries of a number of flowers.

FSH abbreviation for ◊***follicle-stimulating hormone***.

fungus (plural ***fungi***) any of a group of organisms in the kingdom Fungi. Fungi are not considered plants. They lack leaves and roots; they contain no chlorophyll and reproduce by spores. Moulds, yeasts, mildews, and mushrooms are all types of fungus.

Because fungi have no chlorophyll, they must get food from organic substances. They are either ◊parasites, existing on living plants or animals, or ◊saprotrophs, living on dead matter.

G

gall bladder small sac attached to the liver and connected to the small intestine by the ◊bile duct. It stores ◊bile from the liver.

gamete sex cell produced by animals and plants as part of the process of sexual reproduction. The function of the gamete is to carry the genetic message of the parent. During ◊fertilization, two gametes fuse to form a ◊zygote, giving rise to a new and unique individual whose genetic code is slightly different from that of either of the parents. In this way genetic variation persists in a population.

Gametes are ◊haploid, containing only half the number of chromosomes found in the parents. They are produced by ◊meiosis, a particular form of cell division. On fertilization, the chromosomes of each gamete pair up, so that the new individual has the original (◊diploid) number of chromosomes.

ganglion (plural ***ganglia***) a solid cluster of nervous tissue containing many cell bodies and ◊synapses, usually enclosed in a tissue sheath.

gas exchange movement of gases between an organism and the atmosphere, principally oxygen and carbon dioxide. All aerobic organisms (most animals and plants) take in oxygen in order to burn food and manufacture ◊ATP. The resultant reactions release carbon dioxide as a waste product to be passed out into the environment. Green plants also absorb carbon dioxide during ◊photosynthesis, and release oxygen as a waste product.

Specialized ◊respiratory surfaces have evolved during evolution to make gas exchange more efficient. In humans and other tetrapods (four-limbed vertebrates), gas exchange occurs in the ◊lungs, aided by the breathing movements of the ribs. Many adult amphibia and terrestrial invertebrates can absorb oxygen directly through the skin. The bodies of insects and some spiders contain a system of air-filled tubes known as ◊tracheae. Fish have ◊gills as their main respiratory surface. In plants,

gas exchange

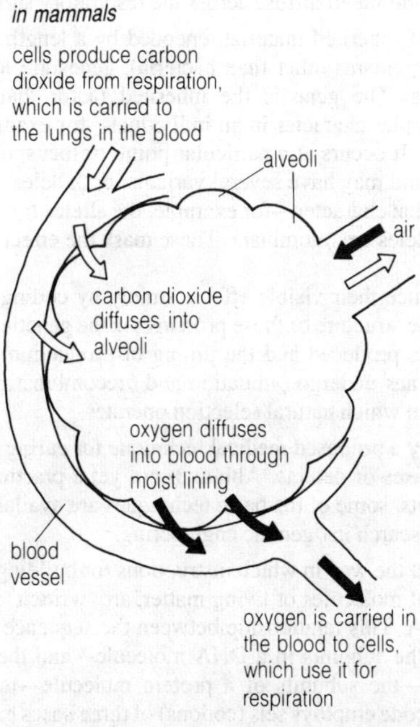

gas exchange generally takes place via the stomata (see ◊stoma) and the air-filled spaces between the cells in the interior of the leaf.

The process of gas exchange relies upon the ◊diffusion of gases across the respiratory surfaces. For example, oxygen diffuses from the lungs into the blood supply because oxygen molecules are at a higher concentration in the alveoli (air sacs) than they are in the capillaries surrounding the alveoli. As the oxygen molecules diffuse along the concentration gradient, they will tend to cross the alveolus and capillary walls and pass into the blood. Gas exchange therefore depends on the ability of the organism to maintain a concentration gradient so that oxygen will continue to diffuse across the respiratory surface.

gene a unit of inherited material, encoded by a length of ◊DNA. In eukaryotes (organisms other than bacteria), genes are located on the ◊chromosomes. The gene is the inherited factor that consistently affects a particular character in an individual—for example, the gene for eye colour. It occurs at a particular point, or locus, on a particular chromosome and may have several variants, or ◊alleles, each specifying a form of that character—for example, the alleles for blue or brown eyes. Some alleles are ◊dominant. These mask the effect of other, ◊recessive alleles.

Genes produce their visible effects simply by coding for proteins; they control the structure of those proteins via the genetic code, as well as the amounts produced and the timing of production (see ◊protein synthesis). Genes undergo ◊mutation and ◊recombination to produce the variation on which natural selection operates.

gene therapy a proposed medical technique for curing or alleviating inherited diseases or defects. Although not yet a practical possibility for most defects, some of the basic techniques are available as a result of intensive research in ◊genetic engineering.

genetic code the way in which instructions for building proteins, the basic structural molecules of living matter, are 'written' in the genetic material ◊DNA. This relationship between the sequence of bases (see ◊base pair)—the subunits in a DNA molecule—and the sequence of ◊amino acids—the subunits of a protein molecule—is the basis of heredity. The code employs sets (codons) of three bases each, and is the same in almost all organisms.

genetic engineering the deliberate manipulation of genetic material by biochemical techniques. It is often achieved by the introduction of new ◊DNA, usually by means of a virus. This can be for pure research or in order to breed new varieties of plants, animals or bacteria for specific purposes—for example, bacteria may be modified so that they secrete rare drugs. Developments in genetic engineering have already led to the industrial production of human insulin, human growth hormone, and a number of vaccines.

genetic fingerprinting technique used for determining the pattern of certain parts of the genetic material ◊DNA that is unique to each individual. The pattern can be determined from a sample of skin, hair, or semen. Like skin fingerprinting, genetic fingerprinting can accurately distinguish humans from one another, with the exception of identical twins. It is used in paternity testing, forensic medicine, and inbreeding studies.

genetics the study of inheritance and of the units of inheritance (◊genes). The founder of genetics was Gregor Mendel, whose experiments with plants, such as peas, showed that inheritance takes place by means of discrete 'particles', which later came to be called genes.

Before Mendel, it had been assumed that the characteristics of the two parents were blended during inheritance, but Mendel showed that the genes remain intact, although their combinations change. Since Mendel, genetics has advanced greatly, first through ◊breeding experiments and optical-microscope observations (classical genetics), later by means of biochemical and electron-microscope studies (molecular genetics). An advance was the elucidation of the structure of ◊DNA by James D Watson and Francis Crick, and the subsequent cracking of the ◊genetic code. These discoveries opened up the possibility of deliberately manipulating genes, or ◊genetic engineering. See also ◊genotype, ◊phenotype, and ◊monohybrid inheritance.

genitalia the reproductive organs of sexually reproducing animals, particulary the external/visible organs of mammals: in males, the penis and the scrotum, which contains the testes, and in females, the clitoris and vulva.

genome the total information carried by the genetic code of a particular organism.

genotype the particular set of ◊alleles (variants of genes) possessed by a given organism. The term is usually used in conjunction with ◊phenotype.

genus (plural ***genera***) in classification, a group of ◊species with many characteristics in common. Thus all doglike species (including dogs, wolves, and jackals) belong to the genus *Canis* (Latin 'dog'). Species of the same genus are thought to be descended from a common ancestor species. Related genera are grouped into ◊families.

geotropism the movement of part of a plant in response to gravity. Roots are positively geotropic because they move downwards towards a gravitational attraction.

germ popular term for a microorganism that causes disease, such as certain ◊bacteria and ◊viruses.

germination the initial stages of growth in a seed, spore, or pollen grain. Seeds germinate when they are exposed to favourable external conditions of moisture, light, and temperature, and when any factors causing dormancy have been removed.

The process begins with the uptake of water by the seed. The embryonic root, or radicle, is normally the first organ to emerge, followed by the embryonic shoot, or plumule. Food reserves, either within the ◊endosperm or from the ◊cotyledons, are broken down to nourish the rapidly growing seedling. Germination is considered to have ended with the production of the first true leaves.

gestation the period from the time of implantation of the embryo in the uterus to birth. This period varies among species; in humans it is about 266 days, in elephants 18–22 months, in cats about 60 days, and in some species of marsupial (such as the opossum) as short as 12 days.

gibberellin plant growth hormone (see also ◊auxin) that mainly promotes stem growth but may also affect the breaking of dormancy in certain buds and seeds, and the induction of flowering.

gill the main respiratory organ of most fishes and immature amphibians, and of many aquatic invertebrates. In all types, water passes over the gills, and oxygen diffuses across the gill membranes into the circulatory system while, carbon dioxide passes from the system out into the water. See ◊gas exchange.

gland a specialized organ of the body that manufactures and secretes enzymes, hormones, or other chemicals. In animals, glands vary in size from the small (for example, tear glands) to the large (for example, the pancreas), but in plants they are always small, and may consist of a single cell. Some glands discharge their products internally like ◊endocrine glands, and others such as ◊exocrine glands, externally.

glomerulus in the kidney, the bundle of blood capillaries responsible for forming the fluid that passes down the tubules and ultimately becomes urine. In the human kidney there are approximately one million tubules, each possessing its own glomerulus.

The structure of the glomerulus allows a wide range of substances including amino acids and sugar, as well as a large volume of water, to pass out of the blood. As the fluid moves through the tubules, most of the water and all of the sugars are reabsorbed, so that only waste remains, dissolved in a relatively small amount of water. This fluid collects in the bladder as urine.

glucagon hormone produced by the pancreas that causes the liver to release sugar into the blood. Its effect is therefore opposite to that of ◊insulin. See also ◊blood glucose regulation.

glucose or ***dextrose*** ($C_6H_{12}O_6$) monosaccharide sugar present in the blood, and found in honey and fruit juices. It is a source of energy for the body, being produced from other sugars and starches to form the 'energy currency' of many biochemical reactions also involving ◊ATP.

glycogen or ***animal starch*** polysaccharide composed of glucose units made and retained in the liver as a carbohydrate store. When required as an energy source by the muscles, it is transported to the muscle tissue where it is converted back to glucose by the hormone ◊insulin and metabolized.

goblet cell a specialized cell occurring in the gut. Goblet cells secrete mucus, a slimy material that covers the wall of the gut and protects it from damage.

Golgi apparatus or ***Golgi body*** a membranous structure found in the cells of ◊eukaryotes. It produces the membranes that surround the cell vesicles or ◊lysosomes.

gonad the part of an animal's body that produces the sperm or ova (eggs) required for sexual reproduction. The sperm-producing gonad is called a ◊testis, and the ovum-producing gonad is called an ◊ovary.

Graafian follicle during the ◊menstrual cycle, a fluid-filled capsule that surrounds and protects the developing egg cell inside the ovary. After the egg cell has been released, the follicle remains and is known as a ◊corpus luteum.

grafting the operation by which a piece of living tissue is removed from one organism and transplanted into the same or a closely related organism where it continues growing. In horticulture, it is a technique widely used for propagating plants, especially woody species. A bud or shoot on one plant, called the ***scion***, is inserted into another, the ***stock***, so that they continue growing together, the tissues combining at the point of union.

grass monocotyledonous plant with about 9,000 species distributed worldwide. The majority are perennial, possessing long, narrow leaves with parallel veins, and jointed, hollow stems; the growing point (meristem) remains close to the ground, growth continuing when the grass is grazed; hermaphroditic flowers are borne in spikelets; the fruits are grainlike. Included in the grass family are wheat, rye, maize, sugarcane, and bamboo.

greenhouse effect a phenomenon of the Earth's atmosphere by which solar radiation, absorbed by the Earth and re-emitted from the surface, is prevented from escaping by carbon dioxide and other gases in the air. The result is a warming of the Earth's atmosphere; in a garden greenhouse, the glass walls have the same effect.

Since the Industrial Revolution and particularly since 1950, levels of carbon dioxide in the atmosphere have risen because of the burning of fossil fuels, causing an intensification of the greenhouse effect and a gradual increase in the Earth's temperature—for example, during the 1980s, the temperature of the world's oceans rose by about 0.1°C a year; the Arctic ice was 6–7 m thick in 1976 and had reduced to 4–5 m by 1987. Increased concentrations of pollutant gases such as the ◊chlorofluorocarbons (CFCs), ozone, and nitrous oxide are also believed to be contributing to global warming. The United Nations Environment Programme estimates that by the year 2025 an increase in average world temperatures will have risen by 1.5°C with a consequent rise of 20 cm in sea level. However, predictions about global warming and its possible climatic effects are tentative, and often conflict.

greenhouse effect

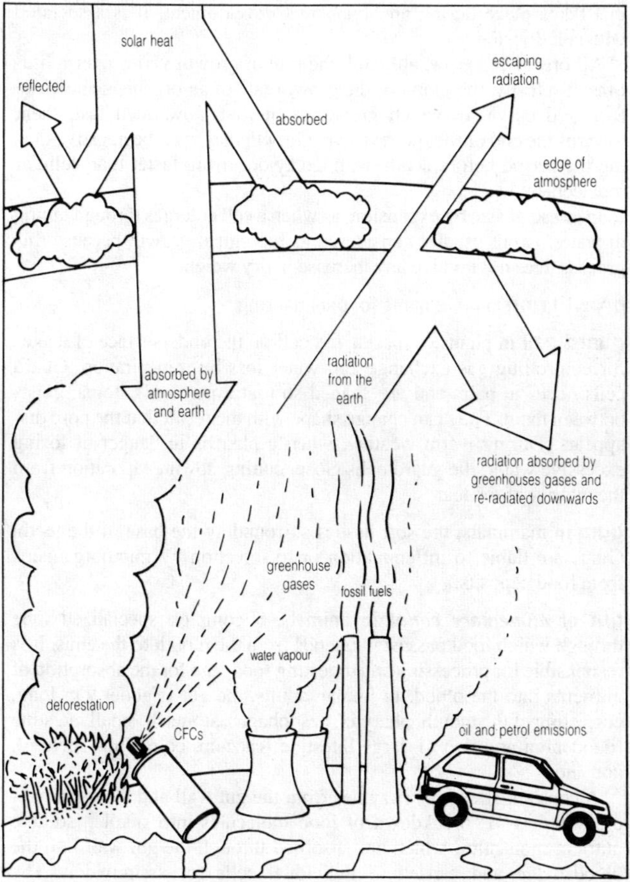

growth the increase in size and dry weight (weight excluding water) that takes place during an organism's development. It is associated with cell division.

All organisms grow, although the rate of growth varies over a lifetime. Typically, the graph of the growth rate of an organism shows an S-shaped curve, in which growth is at first slow, then fast, then, towards the end of life, nonexistent. Growth may even be negative during the period before death, with decay occurring faster than cellular replacement.

Increase of size by expansion, as when a cell enlarges through taking in water, is not usually considered as biological growth because this process does not involve any increase in dry weight.

growth ring another name for ◊annual ring.

guard cell in plants, a specialized cell on the undersurface of leaves for controlling gas exchange and water loss by transpiration. Guard cells occur in pairs and are shaped so that a pore, or ◊stoma, exists between them. They can change shape with the result that the pore disappears. During warm weather, when a plant is in danger of losing excessive water, the guard cells close, cutting down evaporation from the interior of the leaf.

gum in mammals, the soft tissues surrounding the base of the teeth. Gums are liable to inflammation or to infection by microorganisms from food deposits.

gut or ***alimentary canal*** in animals, a complex specialized tube through which food passes; it extends from the mouth to the anus. It is responsible for processing and digesting food, and for the absorption of nutrients into the blood. In human adults, the gut is about 9 m long, consisting of the mouth, pharynx, oesophagus, stomach, small intestine (duodenum and ileum), large intestine (caecum, colon, and rectum), and anus.

A constant stream of enzymes from the gut wall and from the pancreas assists the breakdown of food molecules into smaller, soluble nutrient molecules, which are absorbed through the gut wall into the bloodstream and carried to individual cells for assimilation. The

gut

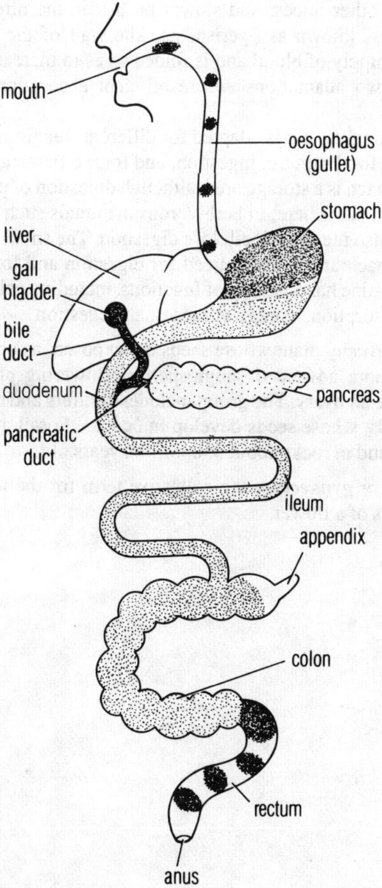
mouth
oesophagus
(gullet)
stomach
liver
gall
bladder
bile
duct
duodenum
pancreas
pancreatic
duct
ileum
appendix
colon
rectum
anus

muscles of the gut keep the incoming food moving, mix it with the enzymes and other juices, and slowly push it in the direction of the anus, a process known as ◊peristalsis. The wall of the gut receives an excellent supply of blood and is folded so as to increase its surface area. These two adaptations ensure efficient absorption of nutrient molecules.

Each region of the gut is adapted for different functions. The mouth is adapted for food capture, ingestion, and for the first stages of digestion. The stomach is a storage area, although digestion of protein by the enzyme pepsin starts here; in herbivorous mammals such as sheep and cattle, this is also the site of cellulose digestion. The small intestine follows the stomach and is specialized for digestion and for absorption. The large intestine has a variety of functions, including cellulose digestion, water absorption, storage of faeces, and egestion.

gymnosperm any plant whose seeds are exposed, as opposed to the structurally more advanced ◊angiosperms (flowering plants), where they are inside an ovary. The group includes conifers and related plants such as cycads, whose seeds develop in ◊cones. Fossil gymnosperms have been found in rocks about 350 million years old.

gynoecium or ***gynaecium*** the collective term for the female reproductive organs of a flower.

H

habitat the localized ◊environment in which an organism lives, and which provides for all (or almost all) of its needs. The diversity of habitats found within the Earth's ecosystem is enormous, and they are changing all the time. Many can be considered inorganic or physical, for example the Arctic ice cap, a cave, or a cliff face. Others are more complex, for instance a woodland or a forest floor. Some habitats are so precise that they are called ***microhabitats***, such as the area under a stone where a particular type of insect lives. Most habitats provide a home for many species.

haemoglobin protein used by all vertebrates and some invertebrates for oxygen transport because the two substances combine reversibly. In vertebrates it occurs in red blood cells (erythrocytes), giving them their colour.

In the lungs or gills where the concentration of oxygen is high, oxygen attaches to haemoglobin forming ***oxyhaemoglobin***. This process

haemoglobin

how oxygen is transported via the blood

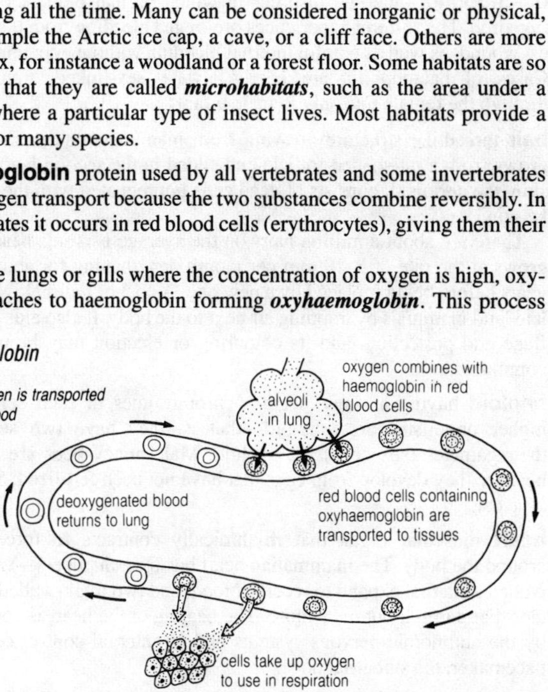

effectively increases the amount of oxygen that can be carried in the bloodstream. The oxygen is later released in the body tissues where it is at low concentration, and the deoxygenated blood returned to the lungs or gills. Haemoglobin will combine also with carbon monoxide, to form carboxyhaemoglobin, but in this case the reaction is irreversible.

haemolysis destruction of red blood cells. Aged cells are constantly being lysed (broken down), but increased wastage of red cells is seen in some infections and blood disorders. It may result in ◊anaemia.

haemophilia inherited disorder in which normal blood clotting is impaired. The sufferer experiences prolonged bleeding from the slightest wound, as well as painful internal bleeding without apparent cause. Cases of haemophilia are nearly always sex-linked, transmitted through the female line only to male infants.

hair threadlike structure growing from mammalian skin. Each hair grows from a pit-shaped follicle embedded in the second layer of the skin, the dermis. It consists of dead cells impregnated with the protein keratin.

There are about a million hairs on the average person's head. Each grows at the rate of 5–10 mm per month, lengthening for about three years before being replaced by a new one. A coat of hair helps to insulate land mammals by trapping air next to the body. It also aids camouflage and protection, and its colouring or erection may be used for communication.

haploid having a single set of ◊chromosomes in each cell. Most higher organisms are ◊diploid—that is, they have two sets—but their gametes (sex cells) are haploid. Male honey bees are haploid because they develop from eggs that have not been fertilized. See also ◊meiosis.

heart muscular organ that rhythmically contracts to force blood around the body. The mammalian heart has four chambers—two thin-walled atria that expand to receive blood, and two thick-walled ventricles that pump it out once more. The beating of the heart is controlled by the autonomic nervous systems and an internal control centre or pacemaker, the sinoatrial node.

heart

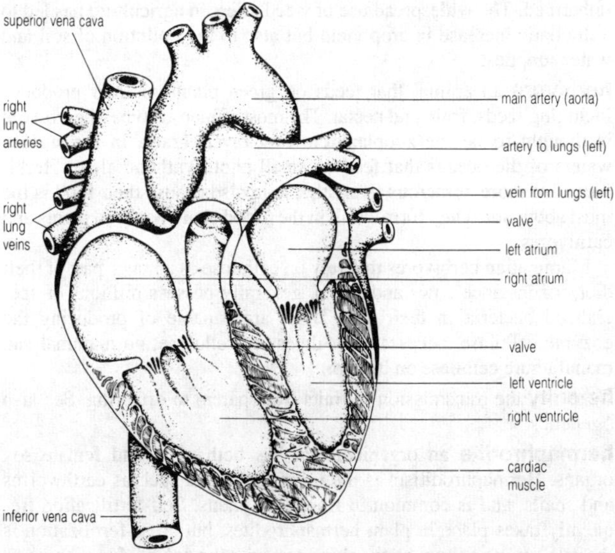

heart beat the regular contraction and relaxation of the heart, and the accompanying sounds. As blood passes through the heart a double beat is heard. The first is produced by the sudden closure of the valves between the atria and the ventricles. The second, slightly delayed sound, is caused by the closure of the valves found at the entrance to the major arteries leaving the heart. Diseased valves may make unusual sounds, known as heart murmurs.

herbaceous plant a plant with very little or no wood, dying back at the end of every summer. The herbaceous perennials survive winters as underground perennating organs such as bulbs and tubers.

herbicide or ***weedkiller*** any chemical used to destroy plants or check their growth. Herbicides may be non-selective, killing all plants, or selective, killing only broad-leaved weeds and leaving cereal crops unharmed. The widespread use of weedkillers in agriculture has led to a dramatic increase in crop yield but also to the pollution of soil and water supplies.

herbivore an animal that feeds on green plants or their products, including seeds, fruit, and nectar. The most numerous type of herbivore is thought to be the zooplankton, tiny invertebrates in the surface waters of the oceans that feed on small photosynthetic algae. Herbivores are more numerous than other animals because their food is the most abundant. They form a link in the ◊food chain between plants and carnivores.

Mammalian herbivores that rely on cellulose as a major part of their diet, for instance cows and sheep, generally possess millions of specialized bacteria in their gut. These are capable of producing the enzyme cellulase, necessary for digesting cellulose; no mammal can manufacture cellulase on its own.

heredity the transmission of traits from parent to offspring. See also ◊genetics.

hermaphrodite an organism that has both male and female sex organs. Hermaphroditism is the norm in species such as earthworms and snails, and is common in flowering plants. Self-fertilization frequently takes place in plant hermaphrodites, but cross-fertilization is the rule among animal hermaphrodites, with the parents functioning as male and female simultaneously, or as one or the other sex at different stages in their development.

heterotroph any living organism that obtains its energy from organic substances produced by other organisms. All animals and fungi are heterotrophs, and they include herbivores, carnivores (including parasites), and saprotrophs (those that feed on dead animal and plant material). Compare ◊autotroph.

heterozygous term describing the possession by an organism of two different ◊alleles for a given character. In ◊homozygous organisms, by contrast, the alleles are identical. In an outbreeding population an individual organism will generally be heterozygous for some genes but homozygous for others.

hibernation a state of ◊dormancy in which certain animals spend the winter. It is associated with a dramatic reduction in all metabolic processes, accompanied by a fall in body temperature, breathing, and heart rate. It is a fallacy that animals sleep throughout the winter.

hinge joint in vertebrates, a ◊joint where movement occurs in one plane only. Examples are the elbow and knee, which are controlled by pairs of muscles, the ◊flexors and ◊extensors.

HIV abbreviation for ***Human Immunodeficiency Virus***, the infectious agent that causes ◊AIDS.

holdfast an organ found at the base of many seaweeds, attaching them to the sea bed. It may be a flattened, sucker-like structure, or dissected and finger-like, growing into rock crevices and firmly anchoring the plant.

holozoic nutrition the type of feeding where an animal ingests lumps of food.

homeostasis the maintenance of a constant state in an organism's internal environment, particularly with regard to pH, salt concentration, temperature, and blood sugar levels. Stable conditions are important for the efficient functioning of the ◊enzyme reactions within the cells, which affect the performance of the entire organism.

homeothermy or ***warm-bloodedness*** maintenance of a constant body temperature in animals, by the use of chemical body processes to compensate for heat loss or gain when external temperatures change. Such processes include generation of heat by the breakdown of food and the contraction of muscles, and loss of heat by sweating, panting, and other means.

Mammals and birds are homeotherms, whereas invertebrates, fish, amphibians, and reptiles are cold-blooded or poikilotherms. Homeotherms generally have a layer of insulating material to retain heat, such as fur, feathers, or fat (see ◊blubber). Their metabolism functions more efficiently due to homeothermy, enabling them to remain active under most climatic conditions.

homologous term describing an organ or structure possessed by members of different taxonomic groups (for example, species, genera, families, orders) that originally derived from the same structure in a common ancestor. The wing of a bat, the arm of a monkey, and the flipper of a seal are homologous because they all derive from the forelimb of an ancestral mammal.

The wing of a bird and the wing of an insect are not homologous, even though they are both used for flying, because they are not derived from the same structure.

homologous chromosomes two chromosomes that possess the same general structure, and, commonly, the same pattern of genes but may have different alleles for those genes. Pairs of homologous chromosomes are found in all diploid organisms, one chromosome from each pair coming from the female parent, the other coming from the male parent. During meiosis, the chromosomes of each pair come together and wrap around each other, exchanging genetic material in a process called ◊recombination; they then separate so that each passes into a different gamete.

homozygous term describing the possession by an organism of two identical ◊alleles for a given character. Individuals homozygous for a character always breed true; that is, they produce offspring that resemble them in appearance when crossed with a genetically similar individual. ◊Recessive alleles are only expressed in the homozygous condition. Compare ◊heterozygous.

honey guides lines or spots on the petals of a flower that indicate to pollinating insects the position of the nectaries (see ◊nectar) within the flower. The orange dot on the lower lip of the toadflax flower is an example. Sometimes the markings reflect only ultraviolet light, which can be seen by many insects although it is not visible to the human eye.

hormone in animals, a substance produced by the ◊endocrine glands, concerned with control of body functions. Hormones are transported in the blood, and bring about changes in the functions of various target organs according to the body's requirements. The pituitary gland, at the base of the brain, is a centre for overall coordination of hormone secretion; the thyroid hormones determine the rate of general body chemistry; the adrenal hormones prepare the organism during stress for 'fight or flight'; and the sex hormones, such as oestrogen, govern reproductive functions.

◊Plant hormones, such as the auxins, are produced in particular regions such as the shoot apex, and regulate growth and development.

host an organism that is parasitized by another. In ◊commensalism, the partner that does not benefit may also be called the host.

human body the physical structure of the human being. It develops from the single cell of the fertilized ovum, is born at 40 weeks, and usually reaches sexual maturity between 11 and 18 years of age. The bony framework (skeleton) consists of more than 200 bones, over half of which are in the hands and feet. Bones are held together by joints, some of which allow movement. The circulatory system supplies muscles and organs with blood, which provides oxygen and food and removes carbon dioxide and other waste products. Body functions are controlled by the nervous system and hormones.

In the upper part of the trunk is the thorax, which contains the lungs and heart. Below this is the abdomen, containing the digestive system (stomach and intestines); the liver, spleen, and pancreas; the urinary system (kidneys, ureters, and bladder); and in women, the reproductive organs (ovaries, uterus, and vagina). In men, the prostate gland and seminal vesicles only of the reproductive system are situated in the abdomen, the testes being in the scrotum, which, with the penis, is suspended in front of and below the abdomen. The bladder empties through a small channel (urethra); in the female this opens in the upper end of the vulval cleft, which also contains the opening of the vagina, or birth canal; in the male, the urethra is continued into the penis. In both sexes, the lower bowel terminates in the anus, a ring of strong muscle situated between the buttocks.

skeleton The skull is mounted on the spinal column, or spine, a chain of 24 vertebrae. The ribs, 12 on each side, are articulated to the spinal column behind, and the upper seven meet the breastbone (sternum) in front. The lower end of the spine rests on the pelvic girdle, composed of the triangular sacrum, to which are attached the hipbones (ilia), which are fused in front. Below the sacrum is the tailbone (coccyx). The shoulder blades (scapulae) are held in place behind the upper ribs by muscles, and connected in front to the breastbone by the two collarbones (clavicles). Each shoulder blade carries a cup (glenoid cavity) into which fits the upper end of the armbone (humerus). This articulates below with the two forearm bones (radius and ulna). These are articulated at the wrist (carpals) to the bones of the hand (metacarpals and phalanges). The upper end of each thighbone (femur) fits into a depression (acetabulum) in the hipbone; its lower end is articulated at the knee to the shinbone (tibia) and calf bone (fibula), which are articulated at the ankle (tarsals) to the bones of the foot (metatarsals and phalanges). At a moving joint, the end of each bone is formed of tough, smooth cartilage, lubricated by ◊synovial fluid. Points of special stress are reinforced by bands of fibrous tissue (ligaments).

Muscles are bundles of fibres wrapped in thin, tough layers of connective tissue (fascia); these are usually prolonged at the ends into strong, white cords (tendons, sinews) or sheets (aponeuroses), which connect the muscles to bones and organs, and by way of which the muscles do their work. Membranes of connective tissue also wrap the organs and line the interior cavities of the body.

The blood vessels of the ***circulatory system***, branching into multitudes of very fine tubes (capillaries), supply all parts of the muscles and organs with blood, which carries oxygen and food necessary for life. The food passes out of the blood to the cells in a clear fluid (lymph); this is returned with waste matter through a system of lymphatic vessels that converge into collecting ducts that drain into large veins in the region of the lower neck. Capillaries join together to form veins which return blood, depleted of oxygen, to the heart.

A finely branching ***nervous system*** regulates the function of the muscles and organs, and makes their needs known to the controlling centres in the central nervous system, which consists of the brain and spinal cord. The inner spaces of the brain and the cord contain cerebrospinal fluid. The body processes are regulated both by the nervous system and by hormones secreted by the endocrine glands.

The thorax has a stout muscular floor, the diaphragm, which expands and contracts the lungs in the act of breathing.

Cavities of the body that open onto the surface are coated with mucous membranes, which secrete a lubricating fluid (mucus). The exterior suface of the body is coated with ***skin***. Within the skin are the sebaceous glands, which secrete sebum, an oily fluid that makes the skin soft and pliable, and the sweat glands, which secrete water and various salts. From the skin grows hair, chiefly on the head, in the armpits, and around the sexual organs; and nails shielding the tips of the fingers and toes; both hair and nail structures are modifications of skin tissue. The skin also contains ◊nerves of touch, pain, heat, and cold.

The human ***digestive system*** is nonspecialized and can break down a wide variety of foodstuffs. Food is mixed with saliva in the mouth by chewing and is swallowed. It enters the stomach, where it is gently churned for some time and mixed with acidic gastric juice. It then passes into the small intestine. In the first part of this, the duodenum, it is broken down further by the juice of the pancreas and duodenal glands, and mixed with bile from the liver, which splits up the fat. The ileum continues the work of digestion and absorbs most of the nutritive substances from the food. The large intestine completes the process, reabsorbing water into the body, and ejecting the useless residue as faeces.

The body, to be healthy, must maintain water and various salts in the right proportions; the process is called ***osmoregulation***. The blood is filtered in the two kidneys, which remove excess water, salts, and metabolic wastes. Together these form urine, which has a yellow pigment derived from bile, and passes down through two fine tubes (ureters) into the bladder, a reservoir from which the urine is emptied at intervals (micturition) through the urethra.

Heat is constantly generated by the combustion of food in the muscles and glands, and by the activity of nerve cells and fibres. It is dissipated through the skin by conduction and evaporation of sweat, through the lungs in the expired air, and in other excreted substances. Average body temperature is about 38°C (37°C in the mouth).

human reproduction an example of ◊sexual reproduction, where the male produces sperm and the female ova (eggs). These gametes contain only half the normal number of chromosomes, 23 instead of 46, so that on fertilization the resulting cell has the correct genetic complement. Fertilization is internal, which increases the chances of conception; unusually for mammals, copulation and pregnancy can occur at any time of the year. Human beings are also remarkable for the length of childhood and for the highly complex systems of parental care found in society. The use of contraception and the development of laboratory methods of insemination and fertilization are issues that make human reproduction more than a merely biological phenomenon.

humerus the upper bone of the forelimb of tetrapods. In humans the humerus is the bone above the elbow.

hybrid the offspring from a cross between individuals of two different species, or two inbred lines within a species. In most cases, hybrids between species are infertile and unable to reproduce sexually.

hybridization the production of a ◊hybrid.

hydrophily a form of ◊pollination in which the pollen is carried by water. It is very rare but occurs in a few aquatic species. In Canadian pondweed, the male flowers break off whole and rise to the water surface where they encounter the female flowers, which are borne on long stalks. In eel grasses, which are coastal plants growing totally submerged, the filamentous pollen grains are released into the water and carried by currents to the female flowers where they become wrapped around the stigmas.

hydroponics the cultivation of plants without soil, using specially prepared solutions of mineral salts.

hypha (plural ***hyphae***) a delicate, usually branching filament, many of which collectively form the mycelium and fruiting bodies of a ◊fungus. Food molecules and other substances are transported along hyphae by the movement of the cytoplasm, known as 'cytoplasmic streaming'.

hypothalamus the region of the brain below the ◊cerebrum which regulates rhythmic activity and physiological stability within the body, including water balance and temperature. It regulates the production of the pituitary gland's hormones and controls that part of the ◊nervous system regulating the involuntary muscles.

I

ileum part of the small intestine of the ◊digestive system, between the duodenum and the colon, that absorbs digested food. Its wall is muscular so that waves of contraction (peristalsis) can mix the food and push it forward. Numerous finger-like projections, or villi (see ◊villus), point inwards from the wall, increasing the surface area available for absorption. The ileum has an excellent blood supply, which receives the food molecules passing through the wall and transports them to the liver via the hepatic portal vein.

immunity the protection that organisms have against foreign microorganisms, such as bacteria and viruses, and against cancerous cells. The cells that provide this protection are called white blood cells, or leucocytes, and make up the immune system. They include neutrophils and ◊macrophages, which can engulf invading organisms and other unwanted material, and natural killer cells that destroy cells infected by viruses and cancerous cells. The lymph nodes play a major role in organizing the immune response.

Immunity is also provided by a range of physical barriers such as the skin, tear fluid, acid in the stomach, and mucus in the airways. ◊AIDS is one of many viral diseases in which the immune system is affected.

implantation in mammals, the process by which the developing ◊embryo attaches itself to the wall of the mother's uterus and stimulates the development of the ◊placenta.

In some species, such as seals and bats, implantation is delayed for several months, during which time the embryo does not grow; thus the interval between mating and birth may be a year, although the ◊gestation period is only seven or eight months.

imprinting the process whereby a young animal learns to recognize both specific individuals (for example, its mother) and its own species. For example, goslings learn to recognize their mother by following the first moving object they see after hatching; as a result, they can easily become imprinted on other species, or even inanimate objects, if these happen to move near them at this time.

inbreeding the mating of closely related individuals. It is considered undesirable because it increases the risk that an offspring will inherit copies of rare deleterious recessive ◊alleles from both parents and so suffer from disabilities.

incisor sharp tooth at the front of the mammalian mouth. Incisors are used for biting or nibbling, as when a rabbit or a sheep eats grass. Rodents, such as rats and squirrels, have large continually-growing incisors, adapted for gnawing. The elephant tusk is a greatly enlarged incisor.

incomplete dominance in genetics, a condition in which a pair of alleles are both expressed in the ◊phenotype because neither can completely mask the other. For example, a red flower crossed with a white flower may produce a pink offspring.

indicator species a plant or animal whose presence or absence in an area indicates certain environmental conditions, such as soil type, high levels of pollution, or, in rivers, low levels of dissolved oxygen. Many plants show a preference for either alkaline or acid soil conditions, while certain trees require aluminium, and are found only in soils where it is present. Some lichens are sensitive to sulphur dioxide in the air, and absence of these species indicates atmospheric pollution.

inflorescence group of ◊flowers, or florets, on a single stem.

ingestion the process of taking food into the mouth. The method of ◊food capture varies but may involve biting, sucking, or filtering. Many single-celled organisms have a region of their cell wall that acts as a mouth. In these cases surrounding tiny hairs (cilia) sweep food particles together, ready for ingestion.

innate behaviour behaviour that does not have to be learnt; see ◊instinct.

inorganic compounds compounds found in organisms that are not typically biological. Water, sodium chloride, and potassium are inorganic compounds because they are widely found outside living cells. The term is also appled to those compounds that do not contain carbon and which are not manufactured by organisms. However, carbon dioxide is considered inorganic, contains carbon, and is manufactured by organisms during respiration. See ◊organic compounds.

insect any member of the class Insecta among the arthropods or jointed-legged animals. An insect's body is divided into head, thorax, and abdomen. The head bears a pair of feelers or antennae, and attached to the thorax are three pairs of legs and usually two pairs of wings. Most insects breathe by means of fine airtubes called tracheae, which open to the exterior by a pair of breathing pores, or spiracles. The young (larvae) of most insects do not resemble the adults and develop by means of a process called ◊metamorphosis. More than one million species are known, and several thousand new ones are discovered every year.

Many insects are seen as pests. They may be controlled by chemical insecticides, by importation of natural predators (biological control), or, more recently, by the use of artificially reared sterile insects, so sharply reducing succeeding generations.

insecticide any chemical pesticide used to kill insects. Among the most effective insecticides are synthetic chlorinated organic chemicals such as DDT; however, these have been shown to accumulate in the environment and to be poisonous to all animal life, including humans, and are consequently banned in many countries. Other synthetic insecticides are now used. Insecticides prepared from plants are safer but need to be applied frequently and carefully.

insectivorous plant a plant that can capture and digest animals, to obtain nitrogen compounds that are lacking in its usual marshy habitat. Some, such as the pitcher plants, are passive traps; others, such as the sundews, butterworts and Venus'-flytrap, have an active trapping mechanism.

instinct behaviour found in all equivalent members of a given species (for example, all the males, or all the females with young) that is presumed to be genetically determined.

Such behaviour is most easily studied in invertebrates such as woodlice and beetles. It is more difficult to identify in vertebrates because it may be combined with or modified by learned behaviour. Instincts differ from ◊reflexes in that they involve very much more complex actions.

insulin hormone, produced by specialized cells in the islets of Langerhans in the ◊pancreas, that regulates the metabolism of glucose, fats, and proteins.

Normally, insulin is secreted in response to rising blood sugar levels (after a meal, for example), stimulating the body's cells to store the excess. Failure of this regulatory mechanism in the disorder *diabetes mellitus*, requires treatment with insulin injections or capsules taken by mouth. Human insulin has now been produced from bacteria by ◊genetic engineering techniques. Compare ◊glucagon.

integument in seed-producing plants, the protective coat surrounding the ovule. In angiosperms (flowering plants) there are two, in gymnosperms only one. A small hole at one end, the micropyle, allows a pollen tube to penetrate through to the egg during fertilization.

intercostal muscle muscle found between the ribs, responsible for producing the rib-cage movements involved in some types of breathing.

When the intercostal muscles contract the ribs move upwards and outwards, enlarging the thorax and causing air to rush into the lungs. On relaxation the ribs move downwards under their own weight, and air is pushed out of the lungs. This type of breathing complements the more gentle contractions of the ◊diaphragm, and occurs for instance during exercise.

intermediate neuron or ***relay neuron*** nerve cell in the spinal cord, connecting motor neurons to sensory neurons. Relay neurons allow information to pass straight through the spinal cord, bypassing the brain. In humans such ◊reflex actions, which are extremely rapid, cause the removal of a limb from a painful stimulus.

intestine in vertebrates, the digestive tract from the stomach outlet to the anus. The human ***small intestine*** is 6 m long, 4 cm in diameter, and consists of the duodenum and ileum; the ***large intestine*** is 1.5 m long, 6 cm in diameter, and includes the caecum, colon, and rectum. Both are muscular tubes comprising an inner lining that secretes alkaline digestive juice, a submucous coat containing fine blood vessels and nerves, a muscular coat, and a serous coat covering all, supported by a strong peritoneum, which carries the blood and lymph vessels, and the nerves. The contents are passed along slowly by ◊peristalsis.

intracellular process series of actions occurring inside a cell. Examples include the intracellular digestion found in some white blood cells, protein synthesis, and duplication of DNA during cell division (mitosis).

intrauterine device IUD or ***coil***, a ◊contraceptive device that is inserted into the uterus (womb). It is a tiny plastic object, sometimes containing copper. By causing a mild inflammation of the lining of the uterus it prevents fertilized eggs from becoming implanted.

IUDs are not usually given to women who have not had children. They are generally very reliable, as long as they stay in place, with a success rate of about 98%. Some women experience heavier and more painful periods, and there is a very small risk of a pelvic infection leading to infertility.

invertebrate an animal without a backbone. The invertebrates comprise over 95% of the million or so existing animal species and include the sponges, coelenterates, flatworms, nematodes, annelid worms, arthropods, molluscs, echinoderms, and primitive aquatic chordates such as sea-squirts and lancelets.

in vivo process an experiment or technique carried out within a living organism.

involuntary action behaviour not under conscious control—for example, the contractions of the gut during peristalsis or the secretion of adrenaline by the adrenal glands. Breathing and urination reflexes are involuntary, although both can be controlled to some extent. These processes are part of the automatic nervous system.

involuntary muscle or ***smooth muscle*** muscle capable of slow contraction over a period of time. Its presence in the gut wall allows slow rhythmic movements known as ◊peristalsis, which cause food to be mixed and forced along the gut. It is also found in the ◊iris of the eye, and plays a part in adjusting the diameter of the pupil in response to changing light intensity.

iris in anatomy, the coloured muscular diaphragm that controls the size of the pupil in the vertebrate eye. It contains longitudinal muscle that increases the pupil diameter and circular muscle that constricts the pupil diameter. Both types of muscle respond involuntarily to light intensity.

irrigation supplying water to dry agricultural areas by means of artificial dams and channels. Irrigation has been practised for thousands of years.

islets of Langerhans group of cells within the pancreas responsible for the secretion of the hormone ◊insulin. They are sensitive to blood-glucose levels, producing more hormone when glucose levels rise.

J

jaw one of two bony structures that form the framework of the mouth in all vertebrates except lampreys and hagfishes. They consist of the upper jawbone (maxilla), which is fused to the skull, and the lower jawbone (mandible), which is hinged at each side to the bones of the temple by ◊ligaments.

joint in vertebrates, the point at which two bones meet.

Some joints allow no motion (the sutures of the skull), others allow a very small motion (the sacroiliac joints in the lower back), but most allow a relatively free motion. Of these, some allow a gliding motion (one vertebra of the spine on another, the bones of the wrist), some have a hinge action (elbow and knee), and others allow motion in all

joint

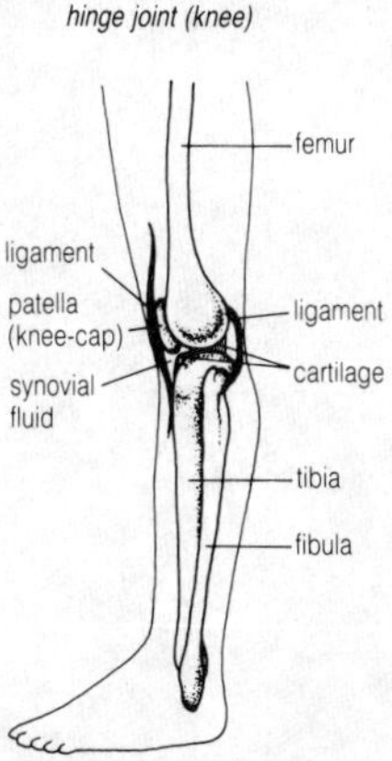

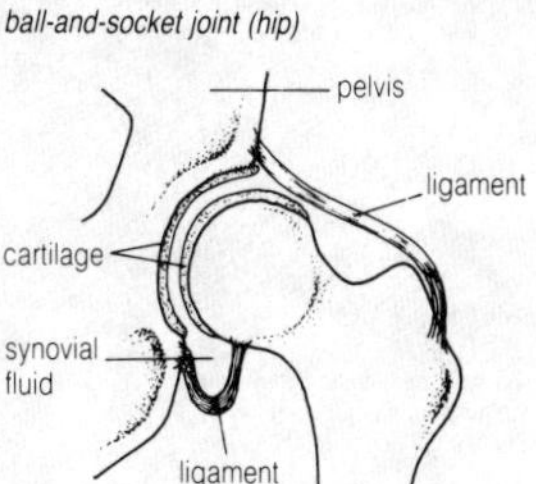

directions (hip and shoulder joints), by means of a ball-and-socket arrangement.

The ends of the bones at a moving joint are covered with cartilage for greater elasticity and smoothness, and enclosed in an envelope (capsule) of tough white fibrous tissue lined with a membrane which secretes a lubricating and cushioning ◊synovial fluid. The joint is further strengthened by ligaments.

jugular vein one of two veins in the necks of vertebrates; they return blood from the head to the superior (or anterior) vena cava and thence to the heart.

K

karyotype the set of ◊chromosomes characteristic of a given species. It is described as the number, shape, and size of the chromosomes in a single cell of an organism. In humans for example, the karyotype consists of 46 chromosomes.

keratin fibrous protein found in the ◊skin of vertebrates and also in hair, nails, claws, hooves, feathers, and the outer coating of horns in animals such as cows and sheep.

If pressure is put on some parts of the skin, more keratin is produced, forming thick callouses that protect the layers of skin beneath.

kernel the inner, softer part of a ◊nut, or of a seed within a hard shell.

key a method of identifying an organism. The investigator is presented with sets of statements, for example 'flower has less than five stamens' and 'flower has five or more stamens'. By successively eliminating statements the investigator moves closer to a positive identification. Identification keys assume a good knowledge of the subject under investigation.

kidney in vertebrates, one of a pair of organs responsible for water regulation, excretion of waste products, and maintaining the ionic composition of the blood. The kidneys are situated on the rear wall of the abdomen. Each one consists of a number of long tubules; the outer parts filter the aqueous components of blood, and the inner parts selectively reabsorb vital salts, leaving waste products in the remaining fluid (urine), which is passed through the ureter to the bladder.

The action of the kidneys is vital, although if one is removed, the other enlarges to take over its function. A patient with two defective kidneys may continue near-normal life with the aid of a kidney machine or continuous ambulatory peritoneal dialysis (CAPD).

kidney

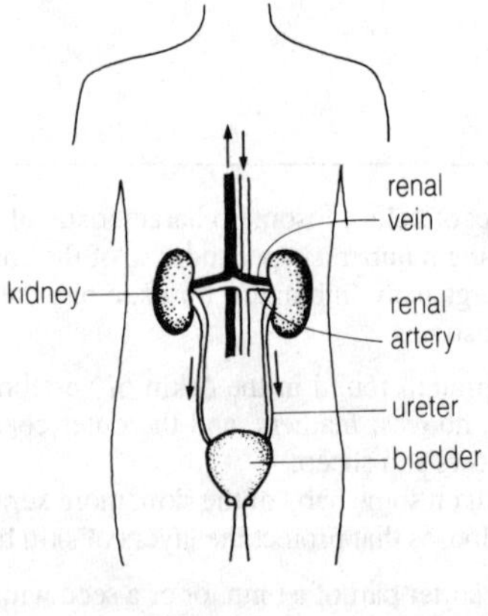

cross section

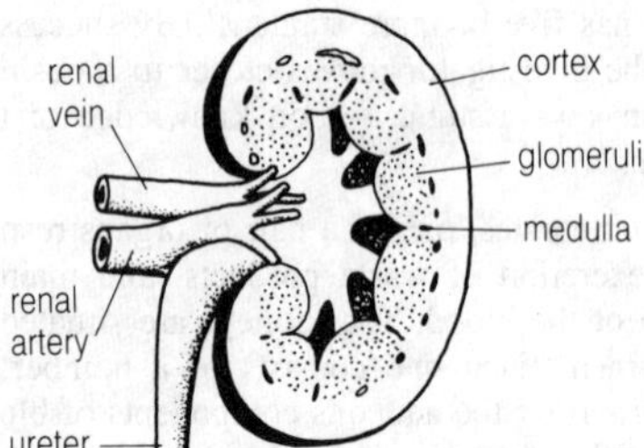

kingdom the primary division in ⟩classification. At one time, only two kingdoms were recognized: animals and plants. Today most biologists prefer a five-kingdom system, even though it still involves grouping together organisms that are probably unrelated. One widely accepted scheme is as follows: ***Kingdom Animalia*** (all multicellular animals); ***Kingdom Plantae*** (all plants, all seaweeds and other algae, including unicellular algae); ***Kingdom Fungi*** (all fungi, including the unicellular yeasts, but not slime moulds); ***Kingdom Protista*** or ***Protoctista*** (protozoa, diatoms, dinoflagellates, slime moulds, and various other lower organisms with eukaryotic cells); and ***Kingdom Monera*** (all prokaryotes—the bacteria and cyanobacteria). The first four of these kingdoms make up the eukaryotes.

Krebs cycle or ***citric acid cycle*** part of the chain of biochemical reactions through which organisms break down food using oxygen (respiration) to release energy. It breaks down food molecules in a series of small steps, producing energy-rich molecules of ◊ATP.

kwashiorkor severe protein deficiency in children under five years, resulting in retarded growth and a swollen abdomen.

L

lactation the secretion of milk from the mammary glands of mammals. In late pregnancy, the cells lining the lobules inside the mammary glands begin extracting substances from the blood to produce milk. The supply of milk starts shortly after birth with the production of colostrum, a clear fluid consisting largely of water, protein, antibodies, and vitamins. The production of milk continues practically as long as the infant continues to suck.

lacteal small vessel responsible for absorbing fat in the small intestine. Occurring in the finger-like villi (see ◊villus) of the ileum, lacteals have a milky appearance and drain into the lymphatic system.

Before fat can pass into the lacteal, bile from the liver causes its ◊emulsification into droplets small enough for attack by the enzyme lipase. The products of this digestion form into even smaller droplets, which diffuse into the villi. Large droplets reform before entering the lacteal and this causes the milky appearance.

lactic acid organic acid produced by certain bacteria during fermentation and by active muscle cells experiencing ◊oxygen debt. It occurs in yoghurt, sour cream, wine, and certain plant extracts. See also ◊anaerobic respiration.

lactose tasteless disaccharide sugar made up of the two monosaccharides glucose and galactose, found in solution in milk; it forms 5% of cow's milk.

Lamarckism a theory of evolution, now discredited, advocated during the early 19th century by the French naturalist Jean Baptiste de Lamarck. It differed from the Darwinian theory of evolution in that it was based on the idea that ◊acquired characters were inherited. For example, he suggested that giraffes have long necks because they are continually stretching them to reach high leaves; according to the theory, giraffes that have lengthened their necks by stretching will pass this characteristic on to their offspring.

lamina in flowering plants (◊angiosperms), the blade of the ◊leaf on either side of the midrib. The lamina is generally thin and flattened, and is usually the primary organ of ◊photosynthesis. It has a network of veins through which water and nutrients are conducted.

large intestine the lower ◊gut or bowels, made up of the colon, the caecum, and the rectum. No absorption of food takes place in the large intestine but the colon removes water from the undigested material, which is then stored as faeces in the rectum.

larva the stage between hatching and adulthood in those species in which the young have a different appearance and way of life from the adults. Examples include tadpoles (frogs) and caterpillars (butterflies and moths). The process whereby the larva changes into another stage, such as a pupa (chrysalis) or adult, is known as ◊metamorphosis.

larynx in mammals, a cavity at the upper end of the trachea (windpipe), containing the vocal cords. It is stiffened with cartilage and lined with mucous membrane.

leaching process by which substances are washed out of the ◊soil. Fertilizers leached out of the soil find their way into rivers and cause water ◊pollution. In tropical areas, leaching of the soil after ◊deforestation removes scarce nutrients and leads to a dramatic loss of soil fertility.

leaf lateral outgrowth on the stem of a plant, and in most species the primary organ of ◊photosynthesis.

Typically leaves are composed of three parts: the sheath or leaf base, the petiole or stalk, and the ◊lamina or blade. The lamina has a network of veins through which water and nutrients are conducted. Structurally the leaf is made up of ◊mesophyll cells surrounded by the epidermis and usually, in addition, a waxy layer, called the ◊cuticle, which prevents excessive evaporation of water from the leaf tissues by ◊transpiration. The epidermis is interrupted by small pores, or stomata (see ◊stoma) through which ◊gas exchange occurs.

leaf

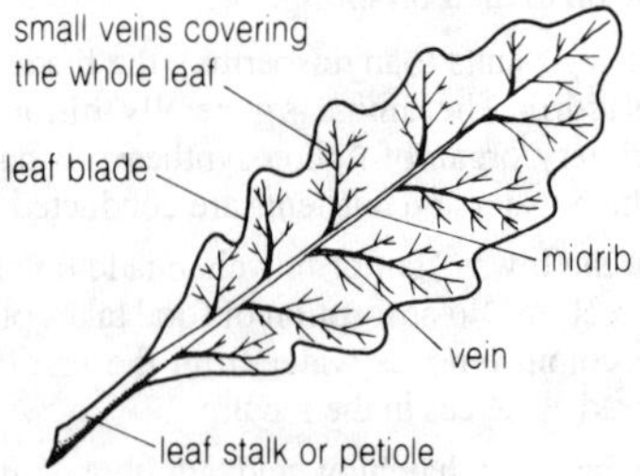

section through a leaf blade

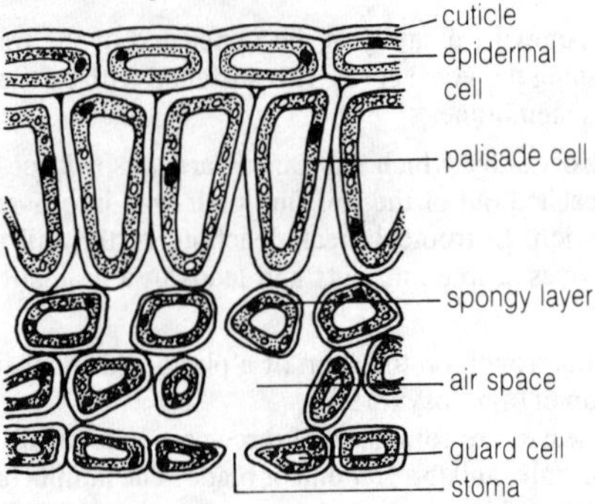

A ***simple leaf*** is undivided, as in the beech or oak. A ***compound leaf*** is composed of several leaflets, as in the blackberry, horsechestnut, or ash tree. Leaves that fall in the autumn are called ***deciduous***, while evergreens are ***persistent***.

learning the changing of an animal's behaviour in response to experience. Although it is associated with the particularly complicated nervous systems of the higher vertebrates, there is evidence to show that even very simple animals are capable of learning.

legume plant belonging to the pea family, which which has a pod containing dry fruits. Legumes are important in agriculture because of their specialized roots, used to fix nitrogen in the soil (see ◊nitrogen fixation).

lenticel a small pore on the stems of woody plants or the trunks of trees. Lenticels are a means of ◊gas exchange between the stem interior and the atmosphere. They consist of loosely packed cells with many air spaces in between, and are easily seen on smooth-barked trees such as cherries, where they form horizontal lines on the trunk.

leucocyte another name for a ◊white blood cell.

LH abbreviation for ***◊luteinizing hormone***.

life the ability to grow, reproduce, and respond to such stimuli as light, heat, and sound. It is thought that life on Earth began about 4,000 million years ago. Over time, life has evolved from primitive single-celled organisms to complex multicellular ones.

Life originated in the oceans. The original atmosphere, 4,000 million years ago, consisted of carbon dioxide, nitrogen, and water. It has been shown in the laboratory that more complex organic molecules, such as ◊amino acids and ◊nucleotides, can be produced from these ingredients by passing electric sparks through a mixture. It has been suggested that lightning was extremely common in the early atmosphere, and that this combination of conditions could have resulted in the oceans becoming rich in organic molecules, the so-called primeval soup. These molecules may then have organized into clusters, capable of reproducing and of developing eventually into simple cells.

life cycle the sequence of developmental stages through which members of a given species pass. Most vertebrates have a simple life cycle consisting of ◊fertilization of sex cells or ◊gametes, a period of development as an ◊embryo, a period of juvenile growth after hatching or birth, an adulthood including ◊sexual reproduction, and finally death.

Invertebrate life cycles are generally more complex and may involve major reconstitution of the individual's appearance (◊metamorphosis) and completely different styles of life. Plants have a special type of life cycle with two distinct phases, known as ◊alternation of generations.

life expectancy the average lifespan of a person at birth. It depends on nutrition, disease control, environmental contaminants, war, stress, and living standards in general.

There is a marked difference between First World countries, which generally have an ageing population and Third World countries, where life expectancy is much shorter. In the UK, average life expectancy for both sexes currently stands at 75; in Bangladesh, it is 48; in Nigeria 49; in famine-prone Ethiopia it is only 41.

lifespan the time between birth and death. Lifespan varies enormously between different species. In humans it is also known as ◊life expectancy and varies according to sex, country, and occupation.

ligament a strong flexible connective tissue, made of the protein collagen, which joins bone to bone at moveable joints. Ligaments prevent bone dislocation (under normal circumstances) but permit joint flexion.

light reaction the first stage of ◊photosynthesis, in which light energy splits water into oxygen and hydrogen ions. The second, dark, stage does not require light and results in the formation of carbohydrates.

lignin a naturally occurring substance produced by plants to strengthen their tissues. It is the essential ingredient of all wood. Lignin is difficult for enzymes to attack, so living organisms cannot digest wood, with the exception of a few specialized fungi and bacteria.

limewater dilute solution of calcium hydroxide used in the laboratory to detect the presence of carbon dioxide. If a gas containing carbon dioxide is bubbled through limewater, the solution turns milky owing to the formation of calcium carbonate. Continued bubbling of the gas causes the limewater to clear again as the calcium carbonate is converted to the more soluble calcium hydrogencarbonate.

limiting factor any factor affecting the rate of a metabolic reaction. Levels of light or of carbon dioxide are limiting factors in ◊photosynthesis because both are necessary for the production of carbohydrates. In experiments, photosynthesis is observed to slow down and eventually stop as the levels of light decrease.

linkage in genetics, the association between two or more genes that tend to be inherited together because they are on the same chromosome. The closer together they are on the chromosome, the less likely they are to be separated by crossing over (one of the processes of ◊recombination) and they are then described as being 'tightly linked'.

lipase the enzyme responsible for breaking down fats into fatty acids and glycerol. It is produced by the ◊pancreas and requires a slightly alkaline environment. The products of fat digestion are absorbed by the intestinal wall.

lipid any of a group of organic compounds soluble in solvents such as ethanol (alcohol), but not in water. They include oils, fats, waxes, and other fatty substances.

liver a large organ of vertebrates, which has many regulatory and storage functions. The human liver is situated in the upper abdomen, and weighs about 2 kg. It receives the products of digestion, converts glucose to glycogen (a long-chain carbohydrate used for storage), and breaks down fats. It removes excess amino acids from the blood, converting them to urea, which is excreted by the kidneys. The liver also synthesizes vitamins, produces bile and blood-clotting factors, and removes damaged red cells and toxins such as alcohol from the blood.

locomotion the ability to move independently from one place to another, occurring in most animals but not in plants. The development of locomotion as a feature of animal life is closely linked to another vital animal feature, that of nutrition. Animals cannot make their food, as can plants, but must find it first; often the food must be captured and killed, which may require great powers of speed. Locomotion is also important in finding a mate, in avoiding predators, and in migrating to favourable areas.

locus the point on a ◊chromosome where a particular ◊gene occurs.

longitudinal muscle type of involuntary muscle found in the gut (alimentary canal) that works antagonistically to ◊circular muscle in peristalsis. It contracts slowly to assist in the mixing of food and in its movement along the gut. Longitudinal muscle is also found in the ◊iris of the eye.

long sight or ***hypermetropia*** eyesight defect where a person is able to focus on objects in the distance, but not on close objects. It is caused by the failure of the lens to return to its normal rounded shape, or by the eyeball being too short, with the result that the image is focused on a point behind the retina. Long-sightedness can be corrected by convex spectacles.

lumen the hollow interior of the gut and the site of digestion in all vertebrates and many invertebrates.

lung large cavity of the body used for ◊gas exchange. It is a specialized ◊respiratory surface, having the form of a large sheet that is folded so as to occupy less space. Most tetrapod (four-limbed) vertebrates have a pair of lungs occupying the thorax. The lung tissue, consisting of multitudes of air sacs and blood vessels, is very light and spongy, and functions by bringing inhaled air into close contact with the blood so that oxygen can be absorbed and waste carbon dioxide can be passed out. The efficiency of lungs is enhanced by breathing movements, by the thinness and moistness of their surfaces, and by a constant supply of circulating blood.

Air is drawn into the lungs through the trachea and bronchi by the expansion of the ribs and the contraction of the diaphragm. The principal diseases of the lungs are bronchitis, emphysema, and cancer (which occur more frequently in smokers than in non-smokers), and pneumonia.

lymph the fluid found in the lymphatic system of vertebrates, which transports nutrients, oxygen, and white blood cells to the tissues, and waste matter away from them. It exudes from ◊capillaries into the tissue spaces between the cells and is made up of blood plasma, plus white cells.

lung

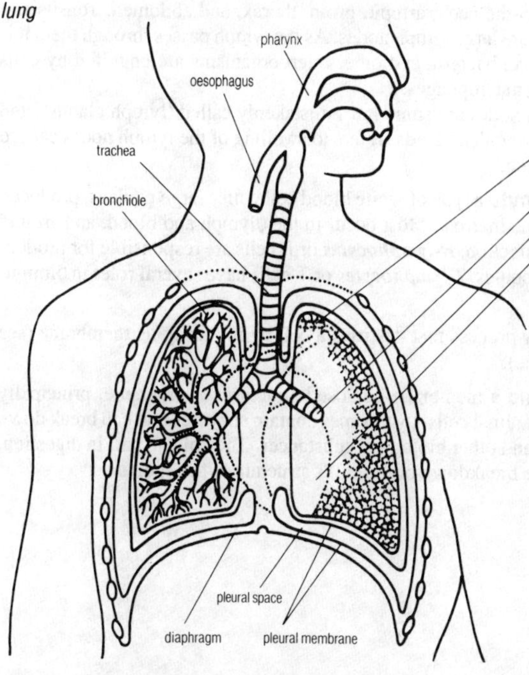

Lymph is drained from the tissues by lymph capillaries, which empty into larger lymph vessels. These lead to ◊lymph nodes, small, round bodies which process the ◊lymphocytes produced by the bone marrow, and filter out harmful substances and bacteria. From the lymph nodes, vessels carry the lymph to the thoracic duct and the right lymphatic duct, which lead into the large veins in the neck.

lymph nodes small masses of lymphatic tissue in the body that occur at various points along the major lymphatic vessels. They are chiefly situated in the neck, armpit, groin, thorax, and abdomen. Tonsils and adenoids are large lymph nodes. As the lymph passes through them it is filtered, and bacteria and other microorganisms are engulfed by cells known as macrophages.

Lymph nodes are sometimes mistakenly called 'lymph glands', and the term 'swollen glands' refers to swelling of the lymph nodes caused by infection.

lymphocyte a type of white blood cell with a large nucleus, produced in the bone marrow. Most occur in the ◊lymph and blood, and around sites of infection. ***B-lymphocytes*** or B cells are responsible for producing ◊antibodies. ***T-lymphocytes*** or T-cells have several roles in ◊immunity.

lysis any process that destroys a cell by rupturing its membrane (see ◊lysosome).

lysosome a membrane-enclosed structure, or organelle, principally found in animal cells. Lysosomes contain enzymes that can break down proteins and other biological substances. They play a part in digestion, and in the breakdown of the cell's structure, when it dies.

M

macrophage a type of ◊white blood cell, or leucocyte, found in all vertebrates. Macrophages specialize in the removal of bacteria and other microorganisms, or of cell debris after injury. Like phagocytes, they engulf foreign matter, but they are larger than phagocytes and have a longer life span. They are found throughout the body, but mainly in the lymph and connective tissues, and especially the lungs, where they ingest dust, fibres, and other inhaled particles.

malnutrition the physical condition resulting from a poor or unbalanced diet. For example, excessive consumption of carbohydrates leads to increasing deposits of fat, and the individual is said to be obese. A lack of protein or of carbohydrate leads to wasting diseases such as kwashiorkor and marasmus, which are eventually fatal. The high global death toll linked to malnutrition arises mostly through diseases killing people weakened by poor diet and an impure water supply.

maltose ($C_{12}H_{22}O_{11}$) dissaccharide sugar made up of two glucose units. It is produced by the action of enzymes on starch and is a major constituent of malt, produced in the early stages of ◊brewing.

mammal any vertebrate that has hair and feeds its young with milk. Mammals are homeothermic (warm-blooded), maintaining a constant body temperature in varied surroundings. Most mammals give birth to live young, but the monotremes, platypus and echidna, lay eggs. There are over 4,000 species, adapted to almost every way of life. The smallest shrew weighs only 2 g, the largest whale up to 150 tonnes.

mammary gland in female mammals, milk-producing gland derived from epithelial cells underlying the skin, active only after the production of young. In all but monotremes (egg-laying mammals), the mammary glands terminate in teats, which aid infant suckling. The number of glands and their position vary between species. In humans there are two (called ◊breasts), in cows four, and in pigs between ten and fourteen.

medulla central part of an organ. The medulla of the kidney lies beneath the outer cortex and is responsible for the reabsorption of water from the filtrate. In the vertebrate brain, it is the posterior region responsible for the coordination of basic activities such as breathing and temperature control. In plants, it is a region of packing tissue in the centre of the stem.

meiosis or ***reduction division*** type of cell division that occurs in the production of ◊gametes, and is therefore important in sexual reproduction. It causes the halving of the chromosome number so that gametes are ◊haploid; this is necessary if the fertilized egg, formed on fusion of the gametes, is to have the same number of chromosomes as the parent. Meiosis is also the stage of the life cycle where genetic variation arises (see ◊recombination); every gamete is slightly different, resulting in nonidentical offspring. See also ◊mitosis.

melanism black coloration of animal bodies caused by large amounts of the pigment melanin. Melanin is of significance in insects, because melanic ones warm more rapidly in sunshine than do pale ones, and can be more active in cool weather. A fall in temperature may stimulate such insects to produce more melanin. In industrial areas, dark insects and pigeons match sooty backgrounds and escape predation, but they are at a disadvantage in rural areas where they do not match their backgrounds. This is known as ***industrial melanism***.

membrane a continuous layer, made up principally of lipid molecules, that encloses a cell or ◊organelles within a cell. Certain small molecules can pass through the cell membrane, but many must enter or leave the cell via channels in the membrane. The ◊Golgi apparatus within the cell is thought to produce certain membranes.

Mendelism in genetics, the theory of inheritance originally outlined by Gregor Mendel. He suggested that, in sexually reproducing species, all characteristics are inherited through indivisible 'factors' (now identifed with ◊genes) contributed by each parent to its offspring. See ◊genetics.

meiosis

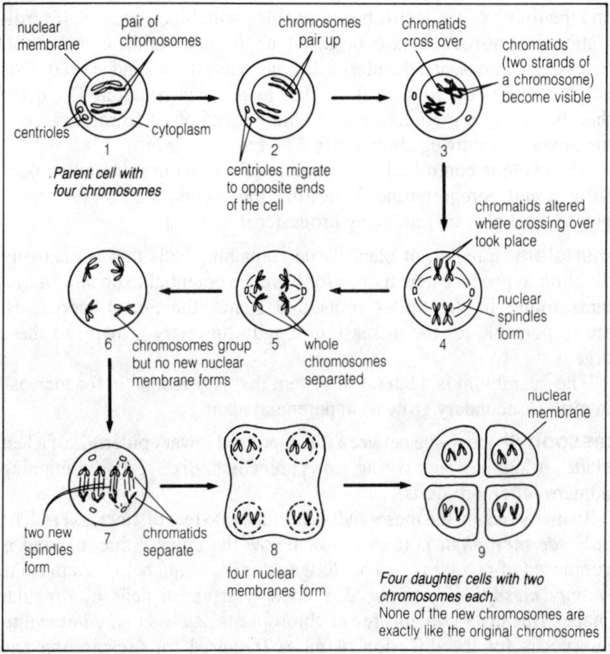

menopause or ***change of life*** in women, the cessation of reproductive ability, characterized by menstruation (see ◊menstrual cycle) becoming irregular and eventually ceasing. The onset is at about the age of 50, but may vary greatly.

menstrual cycle the cycle that occurs in female primates (including humans) of reproductive age, in which the body is prepared for pregnancy. At the beginning of the cycle, a ◊Graafian follicle, nurturing an ovum (egg) within, develops in the ovary, and the inner wall of the uterus forms a soft spongy lining. The ovum is released from the ovary, and the lining of the uterus becomes filled with blood vessels. If fertilization does not occur, the ◊corpus luteum (remains of the Graafian follicle) degenerates, and the uterine lining breaks down, and is shed. This is what causes the loss of blood that marks menstruation. The cycle then begins again. Human menstruation takes place from puberty to menopause, occurring about every 28 days.

The cycle is controlled by a number of ◊hormones, including ◊oestrogen and ◊progesterone. If fertilization occurs, the corpus luteum persists and goes on producing progesterone.

meristem a region of plant tissue containing cells that are actively dividing to produce new tissues (or have the potential to do so). Meristems found in the tip of roots and stems, the apical meristems, are responsible for the increase in length (◊primary growth) of these organs.

The ◊cambium is a lateral meristem that is responsible for increase in girth (◊secondary growth) in perennial plants.

mesophyll the tissue between the upper and lower epidermis of a leaf blade (◊lamina), consisting of parenchyma-like cells containing numerous ◊chloroplasts.

In many plants, the mesophyll is divided into two distinct layers. The ***palisade mesophyll*** is usually just below the upper epidermis and is composed of regular layers of elongated cells. Lying below them is the ***spongy mesophyll***, composed of loosely arranged cells of irregular shape. This layer contains fewer chloroplasts and has many intercellular spaces for the diffusion of gases (required for ◊respiration and ◊photosynthesis), linked to the outside by means of pores called stomata (see ◊stoma).

metabolism the chemical processes of living organisms: a constant alternation of building up (***anabolism***) and breaking down (***catabolism***). For example, green plants build up complex organic substances from water, carbon dioxide, and mineral salts (photosynthesis); by digestion animals partially break down complex organic substances, ingested as food, and subsequently resynthesize them in their own bodies.

metamorphosis a period during the life cycle of many invertebrates, most amphibians, and some fish, during which the individual's body changes from one form to another through a major reconstitution of its tissues. For example, adult frogs are produced by metamorphosis from tadpoles, and butterflies are produced from caterpillars following metamorphosis within a pupa.

The lower orders of insects, such as the dragonflies and the cockroaches, pass through a ***complete metamorphosis***. The young closely resemble the parents and are known as nymphs. However, most insects undergo ***complete metamorphosis***. They hatch at an earlier stage of growth than nymphs and are called larvae. The life of the insect is interrupted by a resting pupal stage when no food is taken. During this stage, the larval organs and tissues are transformed into those of the imago, or adult. Before pupating, the insect protects itself by selecting a suitable hiding place, or making a cocoon of some material which will merge in with its surroundings. When an insect is about to emerge from the pupa, or protective sheath, it undergoes its final moult, which consists of shedding the pupal cuticle.

microbe another name for ◊microorganism.

microbiology the study of organisms that can only be seen under the microscope, mostly viruses and single-celled organisms such as bacteria, protozoa, and yeasts. The practical applications of microbiology are in medicine (since many microorganisms cause disease); in brewing, baking, and other food and beverage processes, where the microorganisms carry out fermentation; and in genetic engineering, which is creating increasing interest in the field of microbiology.

microclimate the particular climate found in a small area or locality.

microorganism or ***microbe*** a living organism invisible to the naked eye but visible under a microscope. Microorganisms include viruses and single-celled organisms such as bacteria, protozoa, yeasts, and some algae. The study of microorganisms is known as microbiology.

micropyle in flowering plants (angiosperms), a small hole towards one end of the ovule. At pollination the pollen tube growing down from the ◊stigma eventually passes through this pore. The male gamete is contained within the tube and is able to travel to the ovum (egg cell) in the interior of the ovule. Fertilization can then take place, with subsequent seed formation and dispersal.

microscope instrument for magnification with high resolution for detail. Optical and electron microscopes are the ones chiefly in use. In 1988 a scanning tunnelling microscope was used to photograph a single protein molecule for the first time.

The ***optical*** or ***light microscope*** usually has two sets of glass lenses and an eyepiece. The ***electron microscope*** passes a beam of electrons, instead of a beam of light, through a specimen and, since electrons are not visible, the eyepiece is replaced with a fluorescent screen or photographic plate; far higher magnification and resolution is possible than with the optical microscope.

migration the movement, either seasonal or as part of a single life cycle, of certain animals, chiefly birds and fish, to distant breeding or feeding grounds.

milk the secretion of the ◊mammary glands of female mammals, with which they suckle their young (during ◊lactation). Over 85% is water, the remainder comprising protein, fat, ◊lactose (a sugar), calcium, phosphorus, iron, and vitamins. The milk of cows, goats, and sheep is often consumed by humans, but only Western societies drink milk after infancy; for people in most of the world, milk causes flatulence and diarrhoea. Milk composition varies among species, depending on the nutritional requirements of the young; human milk contains less protein and more lactose than that of cows.

milk teeth a child's first set of teeth. The complete group of 20 milk teeth are present after two or three years and begin to be replaced by the second permanent set after the age of six. A normal adult dentition consists of 32 teeth, the extra teeth being molars and wisdom teeth.

mimicry the imitation of one species (or group of species) by another. Frequently, the mimic resembles a model that is poisonous or unpleasant to eat, and has warning coloration; the mimic thus benefits from the fact that predators have learned to avoid the model. Hoverflies that resemble bees or wasps are an example. Appearance is usually the

mimicry

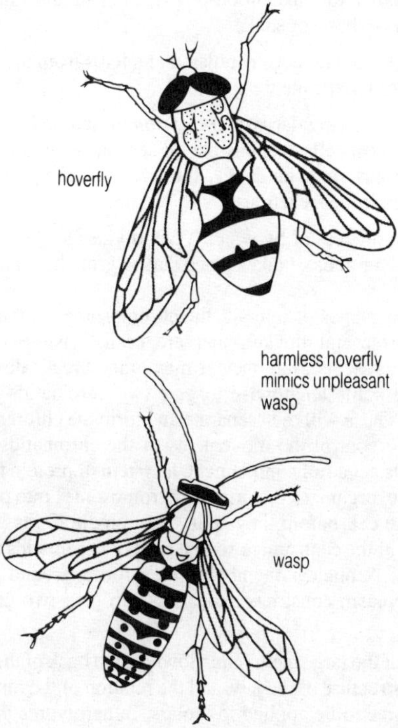

basis for mimicry, but calls, songs, scents, and other signals can also be mimicked.

mineral salt simple inorganic chemical that is required by living organisms. Plants usually obtain their mineral salts from the soil, while animals get theirs from their food. Important mineral salts include iron salts (needed by both plants and animals), magnesium salts (needed mainly by plants, to make chlorophyll), and calcium salts (needed by animals to make bone or shell).

miscarriage spontaneous expulsion of a fetus from the womb before it is capable of independent survival.

mitochondria (singular ***mitochondrion***) small bodies or organelles found inside most cells, responsible for aerobic respiration and for providing the organism with ◊ATP, the essential energy-rich molecule used to drive cellular reactions.

mitosis the process of by which a cell divides to give two identical, diploid daughter cells. It takes place during growth and asexual reproduction.

In the early stages of mitosis, the chromosomes in the cell nucleus become shorter and thicker, and are easily visible when viewed through a microscope; the nuclear membrane breaks down; and each chromosome splits lengthwise to give two chromatids joined at the centromere. These will then separate and form the chromosomes of the new cells. To control the movements of the chromatids so that both daughter cells get a full complement, a system of protein fibres, known as the spindle, organizes the original chromosomes into position in the middle of the cell before they split. The spindle fibres pull the chromatids apart at the centromere so that a set of chromatids forms at each spindle pole. A nuclear membrane then forms around each set and the cell cytoplasm constricts and divides to give two cells. See also ◊meiosis.

molar one of the large teeth found towards the back of the mammalian mouth. The structure of the jaw, and the relation of the muscles, allows a massive force to be applied to molars. In herbivores the molars are flat with sharp ridges of enamel and are used for grinding, an adaptation to a diet of tough plant material. Carnivores have sharp powerful molars called ◊carnassials, which are adapted for cutting meat.

molecular biology the study of the molecular basis of life, including the biochemistry of molecules such as DNA, RNA, and proteins, and the molecular structure and function of the various parts of living cells.

mitosis

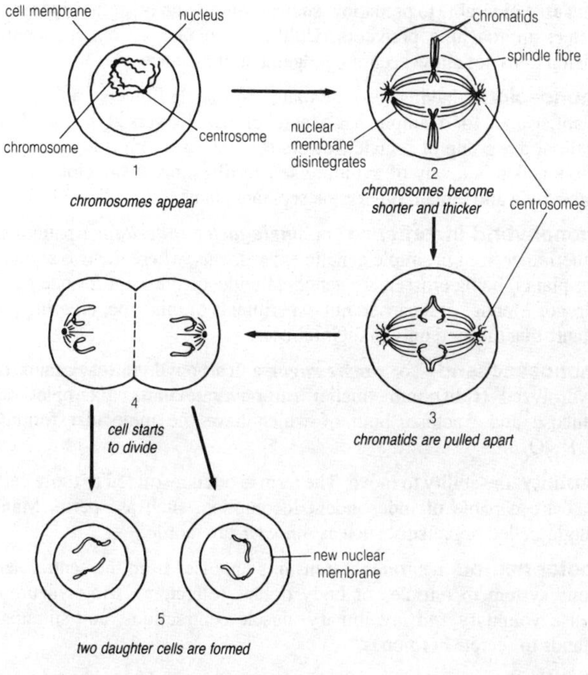

monocotyledon ◊angiosperm (flowering plant) having an embryo with a single cotyledon, or seed leaf (as opposed to ◊dicotyledons, which have two). Monocotyledons usually have narrow leaves with parallel veins and smooth edges, and hollow or soft stems. Their flower parts are arranged in threes. Most are small plants such as grasses, orchids, and lilies, but some are trees such as palms.

monocular vision vision in which the eyes are situated on either side of the head, giving a visual field of almost 360°. It enables animals that are vulnerable to predation, such as rabbits and other herbivores, to detect approaching predators. Unlike ◊binocular vision, monocular vision does not allow accurate judgement of distance.

monoecious having separate male and female flowers on the same plant. Maize, for example, has a tassel of male flowers at the top of the stalk and a group of female flowers (on the ear, or cob) lower down. Monoecism is a way of avoiding self-fertilization. ◊Dioecious plants have male and female flowers on separate plants.

monohybrid inheritance or ***single-factor inheritance*** pattern of inheritance seen in simple genetic experiments, where the two animals (or plants) being crossed are genetically identical except for one gene. Gregor Mendel first carried out experiments of this type, crossing pea plants that differed only in their tallness.

monosaccharide or ***simple sugar*** a ◊carbohydrate that cannot be hydrolyzed (split) into smaller carbohydrate units. Examples are glucose and fructose, both of which have the molecular formula $C_6H_{12}O_6$.

motility the ability to move. The term is often restricted to those cells that are capable of independent locomotion, such as sperm. Many single-celled organisms, such as *Amoeba*, are motile.

motor neuron neuron that transmits impulses from the central nervous system to muscles or body organs (effectors). Motor neurons cause voluntary and involuntary muscle contractions, and stimulate glands to secrete hormones.

mouth the cavity forming the entrance to the gut. In land vertebrates, air from the nostrils enters the mouth cavity to pass down the trachea. The mouth in mammals is enclosed by the jaws, cheeks, and palate.

mucous membrane thin skin lining all those body cavities and canals that come into contact with the air (for example, eyelids, breathing and digestive passages, genital tract). It secretes mucus, a protective fluid.

mucus a lubricating and protective fluid, secreted by mucous membranes in many different parts of the body. In the gut, mucus smooths the passage of food and keeps potentially damaging digestive enzymes away from the gut lining. In the lungs, it traps airborne particles so that they can be expelled.

multicellular organism a creature consisting of many cells. Although single-celled organisms are common, they can only reach a certain size. In the early stages of evolution, increases in size occurred through cells clumping together as colonies. Over the following millions of years organisms became more complex and more efficient by giving different tasks to different groups of cells. In most animals some cells are specialized for nervous control, some for digestion, some for reproduction, and so on.

multiple birth in humans, the production of more than two babies from one pregnancy. Multiple births can be caused by more than two eggs being produced and fertilized (often as the result of hormone therapy to assist pregnancy), or by a single fertilized egg dividing more than once before implantation. See ◊twin.

muscle contractile animal tissue that produces locomotion and maintains the movement of body substances. Muscle is made of long cells that can contract to between one-half and one-third of their relaxed length.

Striped muscles are activated by ◊motor neurons under voluntary control; their ends are usually attached via tendons to bones. ***Involuntary*** or ***smooth*** muscles are controlled by motor neurons of the ◊autonomic nervous system, and located in the gut, blood vessels, iris, and various ducts. ***Cardiac*** muscle occurs only in the heart, and is also controlled by the autonomic nervous system.

mutation change in the genes brought about by a change in the hereditary material ◊DNA. Mutations, the raw material of evolution, result from mistakes during the replication (copying) of DNA molecules. Only a few improve the organism's performance and are therefore favoured by ◊natural selection. Mutation rates are increased by certain chemicals and by radiation.

mutualism or ***symbiosis*** an association between two organisms of different species whereby both profit from the relationship.

mycelium an interwoven mass of threadlike filaments or ◊hyphae, forming the main body of most fungi. The reproductive structures, or 'fruiting bodies', grow from the mycelium.

mycorrhiza a mutually beneficial (mutualistic) association occurring between plant roots and a soil fungus. Mycorrhizal roots take up nutrients more efficiently than non-mycorrhizal roots, and the fungus benefits by obtaining carbohydrates from the tree.

myelin sheath in vertebrates, the fatty, insulating layer that surrounds neurons (nerve cells). It acts to speed up the passage of nerve impulses.

N

nail hard, flat, flexible outgrowth of the digits (fingers and toes) of humans, monkeys, and apes. Nails are made from the same material as hair.

nasal cavity in vertebrates, the organ of the sense of smell; in vertebrates with lungs, it is the upper entrance of the respiratory tract. The cavity opens to the exterior by means of nostrils, which, in most mammals, are set in a protrusion called a ⇨nose. The nasal cavity has a large surface area and is lined with a ⇨mucous membrane that warms and moistens the air and ejects dirt. In the upper parts of the cavity the membrane contains ⇨olfactory cells (cells sensitive to smell).

natural selection process whereby some members of a breeding population are by chance genetically advantaged in relation to the environment, and therefore better able to survive and reproduce than other members. As most environments are slowly but constantly changing, natural selection continually discriminates between members of a population, enhancing the reproductive success of those organisms that possess favourable characteristics. The process is slow, relying on random variation thrown up by ⇨mutation and the genetic ⇨recombination of sexual reproduction; it is believed to be the main cause of ⇨evolution.

nature the living world, including plants, animals, fungi, and all microorganisms, and naturally formed features of the landscape, such as mountains and rivers.

navel a small indentation in the centre of the abdomen of mammals, the remains of the site of attachment of the ⇨umbilical cord, which connects the fetus to the ⇨placenta.

neck the structure between the head and the trunk in animals. In the back of the neck are the upper seven vertebrae, and there are many powerful muscles that support and move the head. In front, the neck region contains the pharynx and trachea, and behind these the oesophagus. The large arteries (carotid, temporal, maxillary) and veins (jugular) that supply the brain and head are also located in the neck.

nectar a sugary liquid secreted by some plants from a nectary, a specialized gland usually situated near the base of the flower. Nectar attracts insects, birds, bats, and other animals to the flower for ◊pollination and is the raw material used by bees in the production of honey.

neo-Darwinism the modern theory of ◊evolution, built up since the 1930s by integrating ◊Darwin's theory of evolution through natural selection with the theory of genetic inheritance founded on the work of Mendel (see ◊Mendelism).

nephron a microscopic unit in vertebrate kidneys that forms ***urine***. A

nephron

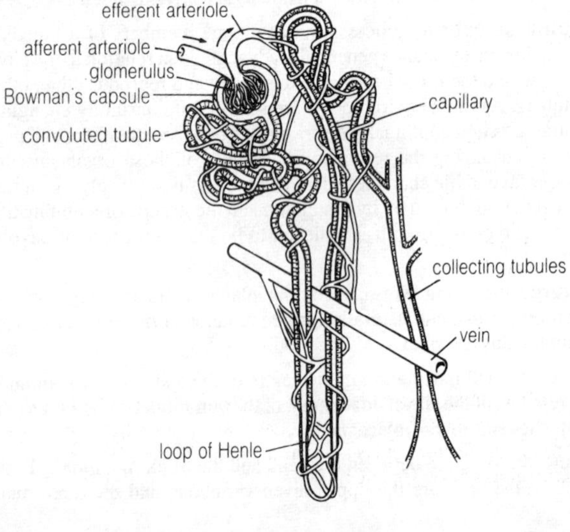

human kidney is composed of over a million nephrons. Each nephron consists of a filter cup surrounding a knot of blood capillaries (glomerulus) and a long narrow collecting tubule in close association with yet more capillaries. Waste materials and water pass from the bloodstream into the filter cup, and essential minerals and some water are reabsorbed from the tubule back into the blood. The urine that is left eventually passes out from the body.

nerve a strand of neurons (nerve cells) enclosed in a sheath of connective tissue joining the ◊central and the ◊autonomic nervous systems with receptor and effector organs. A single nerve may contain both ◊motor neurons and ◊sensory neurons, acting independently.

nervous system the system of interconnected neurons (nerve cells) of all vertebrates and most invertebrates. It is composed of the ◊central and ◊autonomic nervous systems. The mammalian nervous system consists of a central nervous system comprising brain and spinal cord, and a peripheral nervous system connecting up with sensory organs, muscles, and glands.

neuron or ***nerve cell*** type of cell found in the nervous system, capable of rapidly transferring information between different parts of an animal. When neurons are collected together in significant numbers (as in the ◊brain), they not only transfer information but also process it. The unit of information is the ***nerve impulse***, a travelling wave of chemical and electrical changes affecting the membrane of the nerve cell. Many neurons have long extensions, or ◊axons, down which the impulse travels before connecting with another neuron. The impulse involves the passage of sodium and potassium ions across the cell membrane. The axon terminates at the ◊synapse, a specialized area closely linked to the next cell. On reaching the synapse, the impulse releases a chemical (neurotransmitter), which diffuses across to the neighbouring neuron and there stimulates another impulse. See also ◊motor neuron.

neurotransmitter a chemical that diffuses across a ◊synapse, and thus transmits impulses between ◊neurons (nerve cells), or between neurons and effector organs (such as muscles). Nearly 50 different neurotransmitters have been identified.

neuron

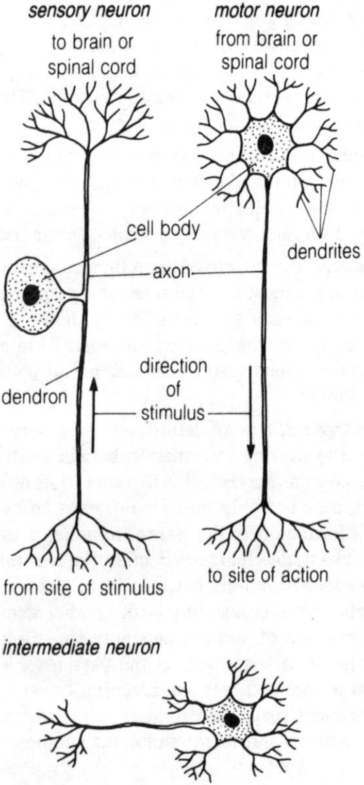

niacin or ***nicotinic acid*** or ***vitamin*** $\boldsymbol{B_3}$ vitamin of the B complex, deficiency of which gives rise to ◊pellagra. Common natural sources are yeast, wheat, and meat. See ◊vitamin.

niche in ecology, the 'place' occupied by a species in its habitat, including all chemical, physical, and biological components, such as what it eats, the time of day at which the species feeds, temperature, moisture, the parts of the habitat that it uses (for example, trees or open grassland), the way it reproduces, and how it behaves. It is believed that no two species can occupy exactly the same niche, because they would be in direct competition for the same resources at every stage of their life cycle.

nitrate inorganic salt containing the NO_3^- ion. Nitrates in the soil, whether naturally occurring or from inorganic (artificial) or organic fertilizers, can be used by plants to make proteins and nucleic acids. Being soluble in water, nitrates are leached out by rain into streams and reservoirs causing ◊eutrophication.

nitrification a process that takes place in soil when bacteria oxidize ammonia, turning it into nitrates. Nitrates can be absorbed by the roots of plants, so this is a vital stage in the ◊nitrogen cycle.

nitrogen cycle in ecology, the process by which nitrogen passes through the ecosystem. Nitrogen, in the form of inorganic compounds (such as nitrates) in the soil, is absorbed by plants and turned into organic compounds (such as proteins) in plant tissue. A proportion of this nitrogen is eaten by ◊herbivores and used for their own biological processes, with some of this in turn being passed on to the carnivores, which feed on the herbivores. The nitrogen is ultimately returned to the soil as excrement and when organisms die and are converted back to inorganic form by bacterial and fungal ◊decomposers.

Although about 78% of the atmosphere is nitrogen, this cannot be used directly by most organisms. However, certain bacteria are capable of ◊nitrogen fixation; that is, they can extract nitrogen directly from the atmosphere and convert it to compounds such as nitrates that other organisms can use. Their presence increases the nitrate content and hence the fertility of the soil.

nitrogen fixation the process by which nitrogen in the atmosphere is converted into nitrates by the action of bacteria and cyanobacteria. Some nitrogen-fixing bacteria live mutually with legumes (peas and

nitrogen cycle

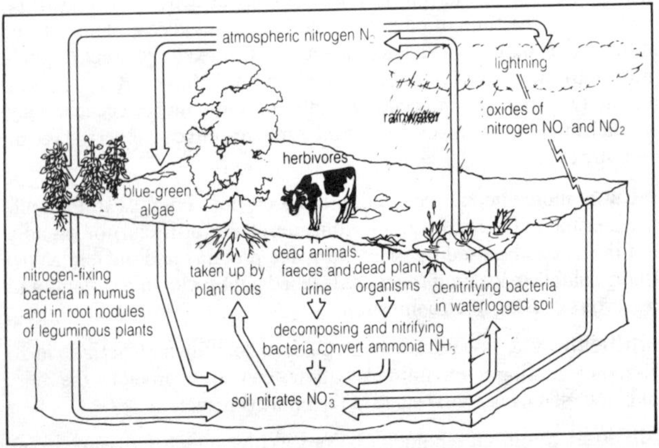

beans) or other plants (for example, alder), where they form nodules on the roots. Such plants are often cultivated in order to increase the fertility of soil. Several chemical processes duplicate nitrogen fixation to produce fertilizers; see ◊nitrogen cycle.

non-reducing sugar disaccharide sugar such as sucrose that does not produce a positive result (change in colour) when tested with Benedict's reagent (see ◊food test). Monosaccharides, such as glucose, and the disaccharides maltose and lactose are ◊reducing sugars.

normal distribution curve the distinctive bell-shaped curve obtained when ◊continuous variation within a population is expressed graphically. When a statistician studies height or intelligence, most people have an intermediate or 'normal' score, with a few individuals scoring either high or low.

nose in humans and most other mammals, a protrusion on the front of the face that contains the nostrils, the external openings of the nasal cavity. It is divided down the middle by a septum of ◊cartilage. The nostrils contain plates of cartilage that can be moved by muscles and have a growth of stiff hairs at the margin to prevent foreign objects from entering.

nostril in vertebrates, the external opening of the ◊nasal cavity. In vertebrates with lungs, the nostrils take in air. In humans, and most other mammals, the nostrils are located on a ◊nose. The posterior passages by which the nasal cavity is connected with the respiratory tract are often called ***internal nostrils***.

nucleic acid complex organic acid made up of a long chain of nucleotides. The two types, known as DNA (deoxyribonucleic acid) and RNA (ribonucleic acid), form the basis of heredity. The nucleotides are made up of a sugar (deoxyribose or ribose), a phosphate group, and one of four purine or pyrimidine bases. The order of the bases along the nucleic acid strand contains the genetic code.

nucleolus dense round structure found in the cell nucleus; it produces the RNA that makes up the ◊ribosomes, following instructions in the DNA.

nucleotide organic compound consisting of a purine (adenine or guanine) or a pyrimidine (thymine, uracil, or cytosine) base linked to a sugar (deoxyribose or ribose) and a phosphate group. ◊DNA and ◊RNA are made up of long chains of nucleotides.

nucleus the central, membrane-enclosed part of a eukaryotic cell (see ◊eukaryote), containing the chromosomes.

nut common name for a dry, single-seeded fruit that does not split open to release the seed. A nut is formed from more than one carpel, but only one seed becomes fully formed; the remainder is aborted. The wall of the fruit, the pericarp, becomes hard and woody, forming the outer shell. Examples are the acorn, hazelnut, and sweet chestnut. The kernels of most nuts provide a concentrated food with about 50% fat and a protein content of 10–20%. Most nuts are produced by ◊perennial trees and bushes.

nutrient any chemical required by an organism in order to live, grow, and reproduce. Animals need complex organic nutrients, whereas plants can take simpler nutrients such as nitrate, and combine them with the products of photosynthesis to make amino acids and proteins, which in turn become the nutrients of animals.

nutrition the stragegy adopted by an organism to obtain the chemicals it needs to live, grow, and reproduce. There are two main types: autotrophic nutrition (plants) and heterotrophic nutrition (animals); see ◊autotroph and ◊heterotroph.

nymph the immature form of insects that do not have a pupal stage—for example, grasshoppers and dragonflies. Nymphs generally resemble the adult (unlike larvae), but do not have fully formed reproductive organs or wings. See ◊metamorphosis.

O

obesity condition of being overweight (generally, 20% or more above the desirable weight for one's sex, build, and height).

Obesity increases susceptibility to disease, strains the vital organs, and lessens life expectancy; it is remedied by healthy diet and exercise, unless caused by glandular problems.

oesophagus the passage by which food travels from mouth to stomach. The human oesophagus is about 23 cm long. Its upper end is at the bottom of the ◊pharynx, immediately behind the trachea.

oestrogen a group of hormones produced by the ◊ovaries of vertebrates; the term is also used for various synthetic hormones that mimic their effects. Oestrogens promote the development of female secondary sexual characteristics; stimulate the production of ova (eggs); and, in mammals, prepare the lining of the uterus for pregnancy.

oestrus in mammals, the period during a female's ◊menstrual cycle when mating is most likely to occur. It usually coincides with ovulation.

oil a mixture of ◊lipids—chiefly triglycerides (lipids containing three ◊fatty acid molecules linked to a molecule of glycerol)—that is liquid at room temperature. See ◊fat.

Mineral oil or petroleum is composed of a mixture of hydrocarbons; in its unrefined form it is a thick greenish-brown liquid found underground in permeable rocks. It is thought to be derived from the fossilized remains of marine organisms laid down millions of years ago.

olfactory cell in mammals, receptor cell found high inside the nasal cavity, associated with the sense of smell. Olfactory cells are stimulated by chemicals in the air and enhance the related sense of taste.

Olfactory cells can be extremely sensitive, although in humans the sense of smell is not well developed. Many mammals rely on the sense of smell for marking out their territories. Insects can respond to minute levels of airborne chemicals such as ◊pheromones.

omnivore an animal that feeds on both plant and animal material. Omnivores have digestive adaptations intermediate between those of ◊herbivores and ◊carnivores, with relatively unspecialized digestive systems and gut microorganisms that can digest a variety of foodstuffs. Examples include the chimpanzee, the cockroach, and the rat.

oosphere another name for the ◊ovum (egg cell) of plants.

optic nerve large nerve passing from the eye to the brain, carrying visual information. In mammals it may contain up to a million sensory neuron fibres, connecting the receptor cells of the retina to the optical centres in the brain.

order in classification, a group of related ◊families. For example, the horse, rhinoceros, and tapir families are grouped in the order Perissodactyla, the odd-toed ungulates, because they all have either one or three toes on each foot. The names of orders are not shown in italic (unlike genus and species names). Related orders are grouped together in a ◊class.

organ part of a living body, such as the liver or brain, that has a distinctive function or set of functions.

organelle a discrete and specialized structure in a living cell; organelles include mitochondria, chloroplasts, lysosomes, ribosomes, and the nucleus.

organic compounds in biochemistry, compounds that contain carbon. They were once thought to be present only in living things; today it is routine to manufacture thousands of organic chemicals both in research and in the drug industry.

organic farming farming without the use of artificial fertilizers (such as ◊nitrates and phosphates) or ◊pesticides (herbicides, insecticides, and fungicides) or other agrochemicals (such as hormones and growth stimulants). Organic farming methods produce food without pesticide residues and greatly reduce pollution of the environment. They are more labour intensive, and therefore more expensive, but use less fossil fuel. Soil structure is greatly improved by organic methods, and recent studies show that a conventional farm can lose four times as much soil through erosion as an organic farm, although the loss may not be immediately obvious.

In place of artificial fertilizers compost, manure, seaweed, or other substances derived from living things are used (hence the name 'organic'). Growing a crop of a nitrogen-fixing plant, such as alfalfa or clover, and then ploughing it back into the soil also fertilizes the ground.

ornithology the study of birds. It covers scientific aspects relating to their structure and classification, and their habits, song, flight, and value to agriculture as destroyers of insect pests.

ornithophily the ◊pollination of flowers by birds. Ornithophilous flowers are typically brightly coloured, often red or orange. They produce large quantities of thin, watery nectar, and are scentless because most birds do not respond well to smell. They are found mostly in tropical areas, with hummingbirds being important pollinators in North and South America, and the sunbirds in Africa and Asia.

osmoregulation the process whereby the water content of living organisms is maintained at a constant level. If the water balance is disrupted, the concentration of salts will be too high or too low, and vital functions, such as nerve conduction, will be adversely affected.

In mammals, loss of water by evaporation is counteracted by increased intake and by mechanisms in the kidneys that enhance the rate at which water is reabsorbed before urine production. Both these responses are mediated by hormones, primarily those of the adrenal cortex (see ◊adrenal gland).

osmosis the movement of solvent such as water through a semipermeable membrane separating solutions of different concentrations. The solvent passes from the more dilute solution to the more concentrated solution until the two concentrations are equal. Many cell membranes behave as semipermeable membranes, and osmosis is a vital mechanism in the transport of fluids in living organisms—for example, in the transport of water from the roots up the stems of plants.

osmosis

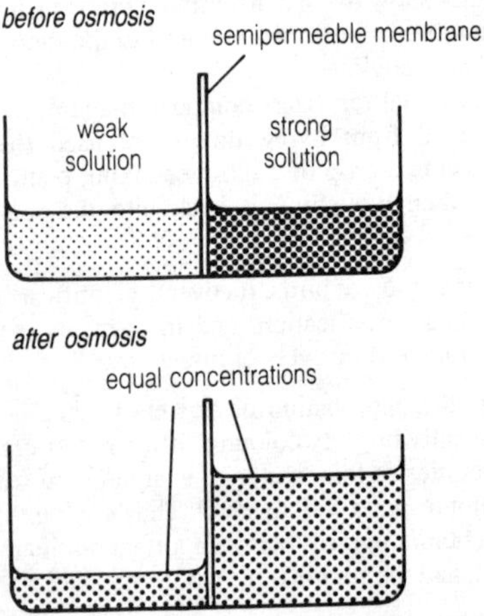

Fish have protective mechanisms to counteract osmosis, which would otherwise cause fluid transport between the body of the animal and the surrounding water (outwards in saltwater fish, inwards in freshwater ones).

osteomalacia softening of the bones, a condition caused by lack of calciferol (vitamin D) in adult life. It results in pain and muscle cramps, bone deformity, and a tendency to spontaneous fracture.

ovary in female animals, the organ that generates the ◊ovum (egg). In humans, the ovaries are two whitish rounded bodies about 25 mm by 35 mm, located in the abdomen near the ends of the ◊Fallopian tubes. Every month, from puberty to the onset of the menopause, an ovum is released from the ovary. This is called ovulation, and forms part of the ◊menstrual cycle. The ovaries secrete the hormones responsible for the secondary sexual characteristics of the female, such as smooth, hairless facial skin and enlarged breasts in humans.

In flowering plants, the ovary is the expanded basal portion of the ◊carpel, containing one or more ◊ovules. It is hollow with a thick wall to protect the ovules. Following fertilization of the ovum within the ovule, it develops into the fruit wall or pericarp.

ovipary a method of animal reproduction in which ◊eggs (fertilized ova) are laid by the female and develop outside her body, in contrast to ovovivipary and vivipary. It is the most common form of reproduction.

ovovivipary a method of animal reproduction in which the eggs (fertilized ova) develop within the female (unlike ovipary), and the embryo gains no direct nourishment from the female (unlike vivipary). It occurs in some invertebrates, fishes, and reptiles.

ovulation in female animals, the process of making and releasing ova (eggs). In humans it occurs as part of the ◊menstrual cycle.

ovule a structure found in angiosperms (flowering plants) and gymnosperms that develops into a seed after fertilization. It consists of an ◊embryo sac containing the female gamete (◊ovum or egg cell), surrounded by nutritive tissue, the nucellus. Outside this there are one or two coverings that provide protection, developing into the testa, or seed coat, following fertilization.

In angiosperms the ovule is within an ◊ovary, but in gymnosperms the ovules are borne on the surface of a scale, usually within a ◊cone, and are not enclosed by an ovary.

ovum (plural ***ova***) the female gamete (sex cell) before fertilization. In animals (where it is also called an egg), it is produced in the ovary; in plants (where it is also known as an egg cell or oosphere), the ovum is produced in an ovule. The ovum is nonmotile. It must be fertilized by a male gamete before it can develop further, except in some cases of ◊parthenogenesis.

oxygen (O_2) colourless odourless gas required by organisms for ◊aerobic respiration and released by green plants as a by-product of ◊photosynthesis. It forms about 21% by volume of the atmosphere.

oxygen debt a physiological state produced by vigorous exercise, in which the lungs cannot supply all the oxygen that the muscles need.

Oxygen is required for the release of energy from food molecules (◊aerobic respiration). Instead of breaking food molecules down fully, muscle cells can switch to a form of partial breakdown that does not require oxygen (◊anaerobic respiration) so that they can continue to generate energy. This partial breakdown produces lactic acid, which results in a sensation of fatigue when it reaches certain levels in the muscles and the blood. Once the vigorous muscle movements cease, the body breaks down the lactic acid, using up extra oxygen to do so. The time required for this is called the ***recovery period***. Panting after exercise is an automatic reaction to 'pay off' the oxygen debt.

oxyhaemoglobin the oxygenated form of ◊haemoglobin, the pigment found in the red blood cells.

ozone (O_3) highly reactive pale-blue gas with a penetrating odour. It forms a layer in the upper atmosphere, which protects life on Earth from ultraviolet rays, a cause of skin cancer. At lower levels it contributes to the ◊greenhouse effect.

A continent-sized hole has formed over Antarctica as a result of damage to the ozone layer caused by ◊chlorofluorocarbons. In 1989 ozone depletion was 50% over the Antarctic compared with 3% over the Arctic. In, 1991 satellite data from NASA revealed that the ozone layer had depleted by 4-8% in the N. hemisphere and by 6-10% in the S. hemisphere between 1978 and 1990. It is believed that the ozone layer is depleting at a rate of about 5% every 10 years over N Europe, with depletion extending south to the Mediterranean and southern USA.

At ground level, ozone can cause asthma attacks, stunted growth in plants, and corrosion of certain materials. It is produced by the action of sunlight on car exhaust fumes, and is a major air pollutant in hot summers.

P

palate in mammals, the ceiling of the mouth. The bony front part is the hard palate, the muscular rear part the soft palate. Incomplete fusion of the two lateral halves of the palate causes interference with speech.

palisade cell cylindrical ◊mesophyll cell lying immediately beneath the upper epidermis of a leaf. Palisade cells normally exist as one closely packed row and contain many chloroplasts. During the hours of daylight palisade cells are photosynthetic, using the energy of the Sun to create carbohydrates from water and carbon dioxide.

pancreas in vertebrates, an accessory gland of the digestive system located close to the duodenum. It secretes enzymes into the duodenum that digest starches, proteins, and fats. In humans, it is about 18 cm long, and lies behind and below the stomach. It contains groups of cells called the ◊***islets of Langerhans***, which secrete the hormones insulin and glucagon that regulate the blood sugar level; see ◊blood glucose regulation.

parasite an organism that lives on or in another organism (called the 'host'), and depends on it for nutrition, often at the expense of the host's welfare. Parasites that live inside the host, such as liver flukes and tapeworms, are called ***endoparasites***; those that live on the outside, such as fleas and lice, are called ***ectoparasites***.

parenchyma a plant tissue composed of loosely packed, more or less spherical cells, with thin cellulose walls. Although parenchyma often has no specialized function, it is usually present in large amounts, forming a packing or ground tissue. It usually has many intercellular spaces.

parental care the time and energy spent by a parent in order to rear its offspring to maturity. Among animals, it ranges from the simple provision of a food supply for the hatching young at the time the eggs are laid (for example, many wasps) to feeding and protection of the young after hatching or birth, as in birds and mammals. In the more social species, parental care may include the teaching of skills—for example, female cats teach their kittens to hunt.

parental generation in genetic crosses, the set of individuals at the start of a test, providing the first set of gametes from which subsequent generations (known as the F1 and F2) will arise.

parthenogenesis the development of an ovum (egg) without any genetic contribution from a male. Parthenogenesis is the normal means of reproduction in a few plants (for example, dandelions) and animals (for example, certain fish). Some sexually reproducing species, such as aphids, show parthenogenesis at some stage in their life cycle.

patella or ***knee cap*** a flat bone embedded in the knee tendon of birds and mammals, which protects the joint from injury.

pathogen microorganism, such as a bacterium or virus, that causes disease. Most pathogens are ◊parasites, and the diseases they cause are incidental to their search for food or shelter inside the host. Nonparasitic organisms, such as soil bacteria or those living in our gut and feeding on waste foodstuffs, can also become pathogenic to a person whose immune system or liver is damaged.

pectoral in vertebrates, the upper area of the thorax associated with the muscles and bones used in moving the arms or forelimbs.

pedicel the stalk of an individual flower, which attaches it to the main stalk of the plant.

pellagra chronic disease of subtropical countries in which the staple food is maize, caused by deficiency of niacin (vitamin B_3), which is contained in protein-rich foods, beans and peas, and yeast. Symptoms include digestive disorders, skin eruptions, and mental disturbances.

pelvis in vertebrates, the lower area of the abdomen featuring the bones and muscles used to move the legs or hindlimbs. The ***pelvic girdle*** is a set of bones that allows movement of the legs in relation to the rest of the body and provides sites for the attachment of relevant muscles.

penis male reproductive organ, used for internal fertilization; it transfers sperm to the female reproductive tract. In mammals, the penis is made erect by vessels that fill with blood. It also contains the urethra, through which urine is passed.

pentadactyl limb the typical limb of tetrapod vertebrates (the mammals, birds, reptiles and amphibians). In basic form, it consists of a 'hand/foot' with five digits (fingers/toes), a lower limb containing two bones, and an upper limb containing one bone. This basic pattern has persisted in all the terrestrial vertebrates, and those aquatic vertebrates (such as seals) which are descended from them. Natural selection has modified the pattern to fit different ways of life. In flying animals (birds and bats) it is greatly altered and in some vertebrates, such as whales and snakes, the limbs are greatly reduced or lost. Pentadactyl limbs of different species are an example of ◊homologous organs.

pepsin enzyme that breaks down proteins during digestion. It requires a strongly acidic environment and is found in the stomach.

peptide a molecule comprising two or more ◊amino acid molecules (not necessarily different) joined by ***peptide bonds***, whereby the acid group of one amino acid is linked to the amino group of the other (–CO.NH). The number of amino amino acid molecules in the peptide is indicated by referring to it as a di-, tri-, or polypeptide (two, three, or many amino acids).

Proteins are built up of interacting polypeptide chains with various types of bonds occurring between the chains.

perennating organ in plants, that part of a ◊biennial plant or herbaceous ◊perennial that allows it to survive the winter; usually a root, tuber, rhizome, bulb, or corm.

perennial plant a plant that lives for more than two years. Herbaceous perennials have aerial stems and leaves that die each autumn. They survive the winter by means of an underground perennating organ, such as a bulb or rhizome. Woody perennials, such as trees and shrubs, have stems that persist above ground throughout the year, and may be either ◊deciduous or ◊evergreen. See also ◊annual plant, ◊biennial plant.

perianth collective term for the outer whorls of the ◊flower, which protect the reproductive parts during development. In most ◊dicotyledons the perianth is composed of two distinct whorls, the calyx of ◊sepals and the corolla of ◊petals, whereas in many ◊monocotyledons they are indistinguishable.

pericarp the wall of a ◊fruit. It encloses the seeds and is derived from the ◊ovary wall. In fruits such as the acorn, the pericarp becomes dry and hard, forming a shell around the seed. In fleshy fruits the pericarp is typically made up of three distinct layers. The ***epicarp***, or ***exocarp***, forms the tough outer skin of the fruits, while the ***mesocarp*** is often fleshy and forms the middle layers. The innermost layer or ***endocarp***, which surrounds the seeds, may be membranous or thick and hard, as in the stone of cherries, plums, and apricots.

period another name for menstruation; see ◊menstrual cycle.

periodontal disease disease of the gums and bone supporting the teeth, caused by the accumulation of ◊plaque and microorganisms; the gums recede, and the teeth eventually become loose and may drop out unless treatment is sought.

peripheral nervous system all parts of the nervous system other than the brain and spinal cord (◊central nervous system).

peristalsis wavelike contractions that pass along tubular organs such as the intestines. They are produced by the alternate contraction and relaxation of antagonistic circular and longitudinal muscles in the organ wall. In the intestines, peristalsis mixes the food and pushes it along.

The same term describes the wavelike motion of earthworms and other invertebrates, in which part of the body contracts as another part elongates.

peritoneum the tissue lining the abdominal cavity and digestive organs.

perspiration or ***sweating*** the excretion of water and dissolved substances from the ◊sweat glands of the skin of mammals. Perspiration has two main functions: body cooling by the evaporation of water from the skin surface, and excretion of waste products such as salts.

pest any insect, fungus, rodent, or other living organism that has a harmful effect on human beings, other than those that directly cause human diseases. Most pests damage crops or livestock, but the term also covers those that damage buildings, destroy food stores, and spread disease.

pesticide any chemical used in farming, gardening, and indoors to combat pests and the diseases they may carry. Pesticides are of three main types: ◊insecticides (to kill insects), ◊herbicides (to kill plants, mainly those considered weeds), and fungicides (to kill fungal diseases).

petal part of a flower whose function is to attract pollinators such as insects or birds. Petals are frequently large and brightly coloured and may also be scented. Some have a nectary at the base and markings on the petal surface, known as ◊honey guides, to direct pollinators to the source of the ◊nectar. In wind-pollinated plants, however, the petals are usually small and insignificant, and sometimes absent altogether. Petals are derived from modified leaves, and are known collectively as the ◊corolla.

petiole stalk attaching the leaf blade, or ◊lamina, to the stem. Typically it is continuous with the midrib of the leaf and attached to the base of the lamina. Leaves that lack a petiole are said to be sessile.

phagocyte a type of ◊white blood cell (leucocyte) that can engulf a bacterium or other invading microorganism. Phagocytes are found in blood, lymph, and other body tissues, where they also ingest foreign matter and dead tissue. A ◊macrophage differs in size and life span.

pharynx the interior of the throat, the cavity at the back of the mouth. Its walls are made of muscle strengthened with a fibrous layer and lined with ◊mucous membrane. The internal nostrils lead backwards into the pharynx, which continues downwards into the oesophagus and (through the epiglottis) into the trachea. On each side, a Eustachian tube enters the pharynx from the middle ear cavity.

phenotype in genetics, the traits actually displayed by an organism. The phenotype is not a direct reflection of the ◊genotype because some alleles are masked by the presence of other, dominant alleles (see

◊dominance). The phenotype is further modified by the effects of the environment (for example, poor food stunting growth).

pheromone chemical signal (such as an odour) that is emitted by one animal and affects the behaviour of others. Pheromones are used by many animal species to attract mates.

phloem a tissue found in vascular plants whose main function is to conduct sugars and other food materials from the leaves, where they are produced, to all other parts of the plant.

Phloem is composed of sieve elements and their associated companion cells, together with some ◊sclerenchyma and ◊parenchyma cell types. Sieve elements are long, thin-walled cells joined end to end, forming sieve tubes; large pores in the end walls allow the continuous passage of nutrients. Phloem is usually found in association with ◊xylem, the water-conducting tissue; unlike the latter it is a living tissue.

photoperiodism mechanism by which the timing of an organism's activities is altered in response to daylength. The flowering of many plants is initiated in this way, and the breeding seasons of many temperate-zone animals are also triggered by increasing or declining day length.

Autumn-flowering plants (for example, chrysanthemum and soya-bean) and autumn-breeding mammals (such as goats and deer) require days that are shorter than a critical length; spring-flowering and spring-breeding ones (such as radish and lettuce; birds) are triggered by longer days.

photosynthesis process by which green plants trap light energy and use it to drive a series of chemical reactions, leading to the formation of carbohydrates. All animals ultimately depend on photosynthesis because it is the method by which the basic food (sugar) is created. For photosynthesis to occur, the plant must possess ◊chlorophyll and must have a supply of carbon dioxide and water. Actively photosynthesizing green plants store excess sugar as starch (this can tested for in the laboratory using iodine).

The chemical reactions of photosynthesis occur in two stages. During the ***light reaction*** sunlight is used to split water (H_2O) into oxygen

photosynthesis

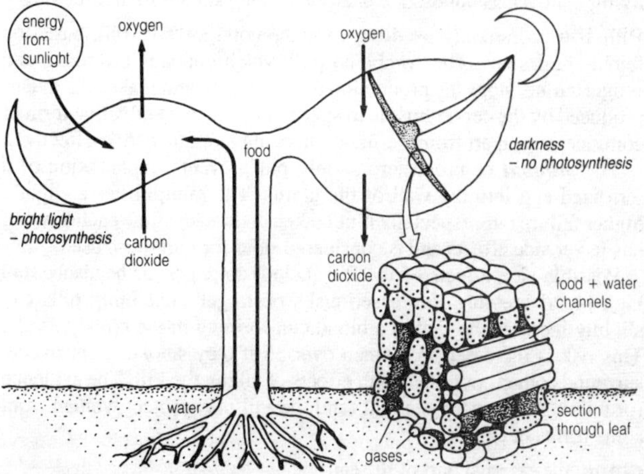

(O_2), protons (hydrogen ions, H^+) and electrons, and oxygen is given off as a by-product. In the second-stage ***dark reaction***, for which sunlight is not required, the protons and electrons are used to convert carbon dioxide (CO_2) into carbohydrates ($C_m(H_2O)_n$).

phototropism movement of part of a plant towards or away from a source of light. Leaves are positively phototropic, detecting the source of light and orientating themselves to receive the maximum amount.

phylum (plural ***phyla***) a major grouping in classification. For example, mammals, birds, reptiles, amphibians, and fishes belong to the phylum Chordata; the phylum Mollusca consists of snails, slugs, mussels, squid, and octopuses; and the phylum Echinodermata includes starfish, sea urchins, and sea cucumbers. All flowering plants belong to a single phylum, Angiospermata, and all conifers to another, Gymnospermata. Related phyla are grouped together in a ◊kingdom; phyla are subdivided into ◊classes.

physiology the branch of biology that deals with the functioning of living animals, as opposed to anatomy, which studies their structures.

Pill, the commonly used term for the contraceptive pill, based on female hormones. The combined pill, which contains oestrogen and progesterone, stops the production of ova (eggs), and makes the mucus produced by the cervix hostile to sperm. It is the most effective form of contraception apart from sterilization, being more than 99% effective.

The ***minipill*** or progesterone-only pill prevents implantation of a fertilized egg into the wall of the uterus. The minipill has a slightly higher failure rate, especially if not taken at the same time each day, but has fewer side effects and is considered safer for long-term use.

Possible side effects of the Pill include migraine or headache and high blood pressure. More seriously, oestrogen-containing pills can slightly increase the risk of a blood clot forming in the blood vessels. This risk is increased in women over 35 if they smoke. Controversy surrounds other possible health effects of taking the Pill. The evidence for a link with cancer is slight (and the Pill may protect women from some forms of cancer).

pinna the external part of the ear.

pioneer species in ecology, those species that are the first to colonize and thrive in new areas. Coal tips, recently cleared woodland, and new roadsides are areas where pioneer species will quickly appear. As the habitat matures other species take over, a process known as ***succession***. See also ◊colonization.

pistil a general term for the female part of a flower, either referring to one single ◊carpel or a group of several fused carpels.

pitfall trap in ecology, a trap consisting of a small container sunk into the earth until its rim is at ground level; invertebrates fall into the container and are unable to escape. The trap is used in surveying the distribution of invertebrates in a particular terrestrial ecosystem.

pituitary gland a major ◊endocrine gland, situated in the centre of the brain. The anterior lobe secretes hormones, some of which control the activities of other glands (thyroid, gonads, and adrenal cortex); others are direct-acting hormones affecting milk secretion, and controlling growth. Secretions of the posterior lobe control body water balance, and contraction of the uterus. The posterior lobe is regulated by nerves from the hypothalamus, and thus forms a link between the nervous and hormonal systems.

placenta in mammals, the organ that attaches the developing ◊fetus to the ◊uterus. It links the blood supply of the fetus to the blood supply of the mother, allowing the exchange of nutrients, wastes, and gases. The two blood systems are not in direct contact, but are separated by thin membranes, with materials diffusing across from one system to the other. The placenta also produces hormones that maintain and regulate pregnancy. It is shed as part of the afterbirth.

It is now understood that a variety of materials, including drugs and viruses, can pass across the placental membrane. HIV, the virus that causes ◊AIDS, can be transmitted in this way.

plant an organism that carries out ◊photosynthesis, has cellulose cell walls and complex eukaryotic cells, and is immobile. A few parasitic plants have lost the ability to photosynthesize but are still considered to be plants.

Plants are ◊autotrophs, that is, they make carbohydrates from water and carbon dioxide, and are the primary producers in all food chains, so that all animal life is dependent on them. They play a vital part in the carbon cycle, removing carbon dioxide from the atmosphere and generating oxygen. The study of plants is known as botany.

Originally the plant kingdom included bacteria and fungi, but these are not now thought of as plants. The groups that are usually classified as plants are the multicellular algae (seaweeds and freshwater weeds), bryophytes (mosses and liverworts), pteridophytes (ferns, horsetails, and club mosses), gymnosperms (conifers, yews, cycads, and ginkgos), and angiosperms (flowering plants).

Many of the lower plants (the algae and bryophytes) consist of a simple body, or thallus, on which the organs of reproduction are borne. They are susceptible to drying out and are therefore usually confined to aquatic or damp habitats. The pteridophytes, gymnosperms, and angiosperms have special supportive water-conducting tissues, which identify them as ◊vascular plants.

plant

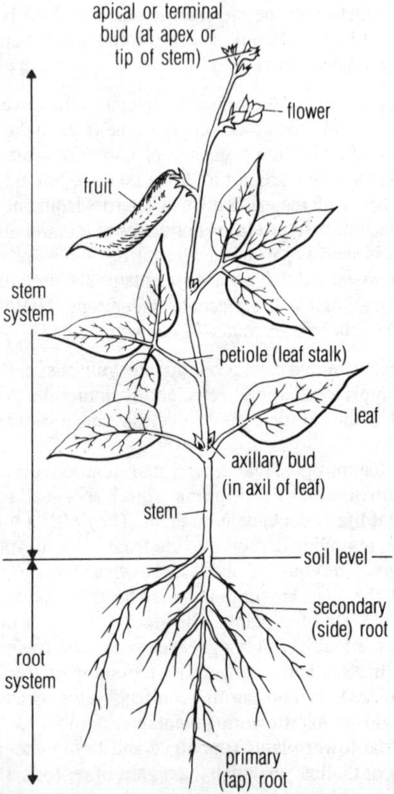

The seed plants (angiosperms and gymnosperms) are the largest group in the plant kingdom, and structurally the most complex. They are usually divided into three parts: root, stem, and leaves. Stems grow above or below ground. Their cellular structure is designed to carry water and salts from the roots to the leaves in the ◊xylem, and sugars from the leaves to the roots in the ◊phloem. The leaves manufacture the food of the plant by means of photosynthesis, which occurs in the ◊chloroplasts they contain. Flowers and cones are modified leaves arranged in groups, enclosing the reproductive organs from which the fruits and seeds result.

plant hormone a substance produced by a plant that has a marked effect on its growth, flowering, leaf fall, fruit ripening, or some other process. Examples include ◊auxin, ◊gibberellin, ethene, and cytokin.

Unlike animal hormones, these substances do not always produce their effect in a different site to that in which they were produced, and they may be less specific in their effects. It has therefore been suggested that they should not be described as hormones at all.

plaque a mixture of saliva, food particles, and bacteria. It builds up on teeth and converts sugars to acid, which eats through the hard exterior enamel into the dentine, causing decay. Plaque can be removed by regular brushing.

plasma the liquid part of the blood.

plasmolysis the separation of the plant cell cytoplasm from the cell wall as a result of water loss. As moisture leaves the vacuole the total volume of the cytoplasm decreases while the cell itself, being rigid, hardly changes. Plasmolysis is induced in the laboratory by immersing a plant cell in a strongly saline or sugary solution, so that water is lost by osmosis. Plasmolysis is unlikely to occur in the wild except in severe conditions.

platelet or ***thrombocyte*** a tiny 'cell' found in the blood, which helps it to clot. Platelets are not true cells, but membrane-bound cell fragments that bud off from large cells in the bone marrow. See ◊blood clotting.

pleural membrane one of a pair of membranes surrounding the lungs, protecting and lubricating them during breathing movements.

plumule the part of a seed embryo that develops into the shoot, bearing the first true leaves of the plant.

poikilothermy or ***cold-bloodedness*** the condition in which an animal's body temperature is largely dependent on the temperature of the air or water in which it lives. It is characteristic of all animals except birds and mammals, which maintain their body temperatures by ◊homeothermy. Poikilotherms have some means of warming themselves up, such as basking in the sun, or shivering, and can cool themselves down by sheltering from the Sun under a rock or by bathing in water.

Poikilotherms are often referred to as 'cold-blooded animals', but this is not really correct: their blood may be as warm as their surroundings, which means it may be warmer than the blood of birds and mammals, for example, in very hot climates.

pollen in angiosperms (flowering plants) and gymnosperms, the grains that contain the male gametes. In angiosperms pollen is produced within ◊anthers; in most ◊gymnosperms it is produced in male ◊cones. A pollen grain is typically yellow and, when mature, has a hard outer wall. Pollen of insect-pollinated plants (see ◊pollination) is often sticky and spiny, and larger than the smooth, light grains produced by wind-pollinated species.

pollination the process by which fertilization occurs in the sexual reproduction of higher plants. The male gametes are contained in pollen grains, which must be transferred from the ◊anther to the ◊stigma in angiosperms (flowering plants), and from the male cone to the female cone in gymnosperms.

Self-pollination occurs when pollen is transferred to a stigma of the same flower, or to another flower on the same plant; ***cross-pollination*** occurs when pollen is transferred to another plant. This involves external pollen-carrying agents, such as wind, water, and animals such as insects, birds, bats, and other small mammals.

Most flowers are adapted for pollination by one particular agent only. Those that rely on animals are generally scented, and have large, brightly coloured petals and sticky pollen grains. They produce nectar, a sugary liquid, or surplus pollen, or both, on which the pollinator feeds. Thus, the relationship between pollinator and plant is an example of mutualism, in which both benefit. In some plants, however, the pollinator receives no benefit, while in others, nectar may be removed by animals that do not bring about pollination.

Wind-pollinated, or anemophilous, flowers are usually unscented, have either very reduced petals and sepals or lack them altogether, and do not produce nectar. Their pollen grains are small and smooth walled, and, because air movements are random, are produced in vast amounts. The male flowers have numerous exposed stamens, often on long filaments; the female flowers have long, often branched, feathery stigmas. Many wind-pollinated plants, such as hazel and birch, bear their flowers on catkins in order to facilitate the free transport of pollen.

pollution the damaging effect on the environment of those substances that human populations produce, principally through industrial and agricultural processes. Pollution can alter the balance of ◊ecosystems, making it difficult or impossible for plants and animals to thrive and reproduce, and can endanger human health, causing birth defects and diseases such as cancer. It has become the central issue for environmental groups such as Friends of the Earth, who have campaigned for stricter laws on pollution control.

vehicle emissions Cars and lorries release a range of pollutants into the atmosphere. Carbon monoxide reduces breathing efficiency by combining irreversibly with ◊haemoglobin in the blood; oxides of nitrogen produce ◊acid rain and can aggravate respiratory conditions such as asthma; particles of carbon react in sunlight to produce low-level ozone, which also causes respiratory problems; and lead compounds impair the development of children. In addition, motor vehicles release large amounts of carbon dioxide, which contributes to the ◊greenhouse effect.

industrial pollution All industry produces waste of some sort; this becomes pollution if not properly managed. For example, most industrial processes involve the use of electricity. However, almost all power stations, in generating electricity, produce large volumes of warm water, which can alter river ecosystems if released without sufficient cooling. Power stations that burn fossil fuels inevitably release large amounts of carbon dioxide into the air—the smoke produced by coal-fired power stations also contains sulphur dioxide, a major cause of acid rain—and nuclear power stations produce radioactive waste, which may remain hazardous for tens of thousands of years.

agricultural pollution Intensive farming requires that land be used year after year without interruption. Fast-acting artificial fertilizers, containing high concentrations of nitrates, are therefore necessary. The manufacture of fertilizers requires a considerable amount of electricity, and is therefore in itself an indirect cause of pollution. Once put on the fields, nitrates can pass into rivers and streams, causing ◊eutrophication, and into drinking water, posing a possible hazard to health. High-yield agriculture also makes use of large quantities of herbicides and insecticides in order to prevent crop damage. Such chemicals can easily pass into the ◊food chain, eventually becoming concentrated in the tissues of predators, causing their death or impairing their ability to reproduce.

polypeptide a long-chain ◊peptide.

polysaccharide a long-chain ◊carbohydrate made up of hundreds or thousands of linked simple sugars (monosaccharides) such as glucose and closely related molecules.

The polysaccharides are natural polymers. They either act as energy-rich food stores in plants (starch) and animals (glycogen), or have structural roles in the plant cell wall (cellulose, pectin) or the tough outer skeleton of insects and similar creatures (chitin). See also ◊carbohydrate.

polyunsaturate a type of triglyceride (◊fat or oil) in which the long carbon chains of the ◊fatty acids contain several double bonds. By contrast, the carbon chains of saturated fats (such as lard) contain only single bonds.

The more double bonds the chains contain, the lower the melting point of the triglyceride. Unsaturated chains with several double bonds produce oils, such as vegetable and fish oils, which are liquids at room temperature. Saturated fats, with no double bonds, are solids at room temperature. The polyunsaturated fats used for margarines are produced by taking a vegetable or fish oil and turning some of the double bonds to single bonds, so that the product is semi-solid at room temperature. Medical evidence suggests that polyunsaturated fats are less likely to contribute to heart disease and arteriosclerosis than are saturated fats.

population cycle regular fluctuations in the size of a population, as seen in lemmings, for example. Such cycles are often caused by density-dependent mortality: high mortality due to overcrowding causes a sudden decline in the population, which then gradually builds up again. Population cycles may also result from an interaction between a predator and its prey.

population genetics the branch of genetics that studies the way in which the frequencies of different ◊alleles in populations of organisms change, as a result of natural selection and other processes.

predator an animal that must hunt and kill other animals for its food.

pregnancy in humans, the period during which an embryo grows within the womb. It begins at conception and ends at birth, and the normal length is 40 weeks. Menstruation usually stops on conception. Pregnancy in animals is called ◊gestation.

premolar in mammals, one of the large teeth towards the back of the mouth. In herbivores they are adapted for grinding. In carnivores they may be ◊carnassials. Premolars are present in milk dentition (see ◊milk teeth) as well as permanent dentition.

preservative food ◊additive used to inhibit the growth of bacteria, yeasts, mould, and other microorganisms to extend the shelf-life of foods. All preservatives are potentially damaging to health if eaten in sufficient quantity. Both the amount used, and the foods in which they can be used, are restricted by law. See also ◊food technology.

primary growth in plants, an increase in size due to cell division and cell expansion. It is restricted to the ◊meristems, those areas of the plant that contain actively dividing cells. The main result of primary growth is the lengthening of the stem and the root.

The type of growth that produces a thickening of the stem involves different tissues and is known as ***secondary growth***.

primary sexual characteristic the endocrine gland producing maleness and femaleness. In males it is the testis and in females the ovary.

pregnancy

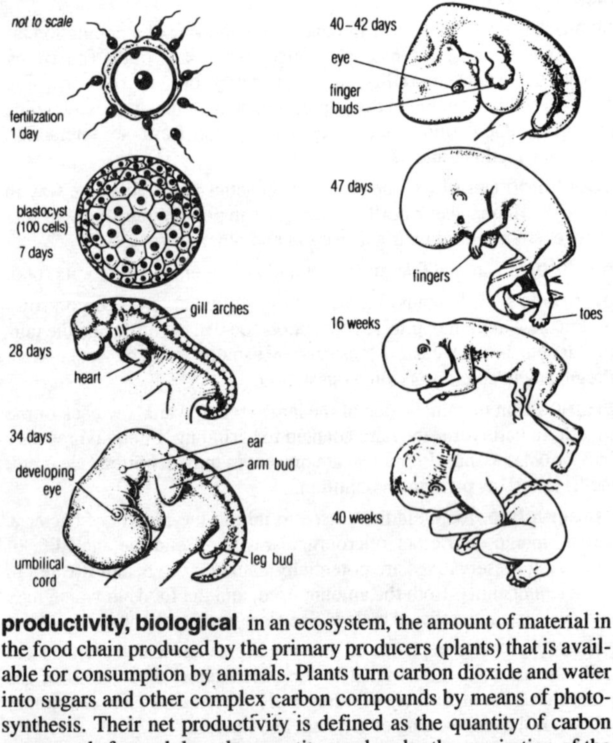

productivity, biological in an ecosystem, the amount of material in the food chain produced by the primary producers (plants) that is available for consumption by animals. Plants turn carbon dioxide and water into sugars and other complex carbon compounds by means of photosynthesis. Their net productivity is defined as the quantity of carbon compounds formed, less the quantity used up by the respiration of the plant itself.

progesterone hormone that regulates the menstrual cycle and pregnancy. It is secreted by the ◊corpus luteum (the ruptured ◊Graafian follicle of a discharged ovum).

prokaryote organism whose cells lack organelles (specialized segregated structures such as nuclei, mitochondria, and chloroplasts). Prokaryote DNA is not arranged in chromosomes but forms a coiled structure called a ***nucleoid***. The prokaryotes comprise only the ***bacteria*** and ***cyanobacteria***; all other organisms are eukaryotes.

propagation of plants in horticulture and agriculture, method used in order to bring about the ◊vegetative reproduction of plants. Its advantage is that all the offspring (daughter plants) are identical; the gardener therefore has control over the features that they show. The various propagation techniques all involve taking tissue from one plant and growing it in another area. Traditional methods include the taking of ◊cuttings, the ◊grafting of shoots from one plant onto the roots of another, and allowing a plant such as the strawberry to grow ◊runners. Modern techniques include the removal of just a few cells from the parent, which can then be grown in a nutrient medium until they form the mature plant.

prostate gland the gland surrounding, and opening into, the urethra at the base of the bladder in male mammals. The prostate gland produces an alkaline fluid that is released during ejaculation; this fluid activates sperm, and prevents their clumping together.

protandry in a flower, the state where the male reproductive organs reach maturity before those of the female. This is a common method of avoiding self-fertilization. See also ◊protogyny.

protease general term for an enzyme capable of splitting proteins. Examples include pepsin, found in the stomach, and trypsin, found in the small intestine.

protein long-chain molecule composed of amino acids joined by ◊peptide bonds. Proteins are essential to all living organisms. As ***enzymes*** they regulate all aspects of metabolism. Structural proteins such as ***keratin*** and ***collagen*** make up the skin, claws, bones, tendons, and ligaments; ***muscle*** proteins produce movement; ***haemoglobin*** transports oxygen; and ***membrane proteins*** regulate the movement of substances into and out of cells.

For humans, protein is an essential part of the diet, and is found in greatest quantity in soya beans and other legumes, meat, eggs, and cheese.

protein synthesis the manufacture, within the cytoplasm of the cell, of the ◊proteins an organism needs. The building blocks of proteins are ◊amino acids, of which their are 21 types. The pattern in which the amino acids are linked decides what kind of protein is produced. In turn it is the genetic code or ◊DNA that determines the precise order in which the amino acids are linked up during protein manufacture. Interestingly, DNA is found only in the nucleus, yet protein synthesis only occurs in the cytoplasm. The information necessary for making the proteins is carried from the nucleus to the cytoplasm by another nucleic acid, ◊RNA.

prothrombin precursor enzyme involved in ◊blood clotting. When a blood vessel is damaged, prothrombin is converted to the active enzyme thrombin, which brings about the formation of fibrin, the insoluble protein that forms a mesh over the wound.

protogyny in a flower, the state where the female reproductive organs reach maturity before those of the male. Like ◊protandry, this is a method of avoiding self-fertilization, but it is much less common.

protoplasm the contents of a living cell. Strictly speaking it includes all the discrete structures (organelles) in a cell, but it is often used simply to mean the jelly-like material in which these float. The contents of a cell outside the nucleus are called ◊cytoplasm.

protozoa a group of single-celled organisms without rigid cell walls. Some, such as *Amoeba*, ingest other cells, but most are ◊saprotrophs or parasites.

pseudocarp a fruit-like structure that incorporates tissue that is not derived from the ovary wall. The additional tissues may be derived from floral parts such as the ◊receptacle and ◊calyx. For example, the coloured, fleshy part of a strawberry develops from the receptacle and the true fruits are the small 'pips' embedded in its outer surface. Pineapples, figs, apples, and pears are pseudocarps.

puberty stage in human development when the individual becomes sexually mature. It may occur from the age of ten upwards. The sexual organs take on their adult form and pubic hair grows. In girls, menstruation begins, and the breasts develop; in boys, the voice breaks and becomes deeper, and facial hair develops.

pulse impulse transmitted by the heartbeat throughout the arterial system. When the heart muscle contracts, it forces blood into the ◊aorta. Because the arteries are elastic, the sudden rise of pressure causes a throb or sudden swelling through them. In humans, the actual flow of the blood is about 60 cm a second; the pulse rate is generally about 70 per minute. The pulse can be felt where an artery is near the surface, such as in the wrist or the neck.

pupa the nonfeeding, largely immobile stage of some insect life cycles, in which larval tissues are broken down, and adult tissues and structures are formed.

In butterflies and moths, the pupa is called a chrysalis.

pure breeding line in genetics, a strain of individuals that when interbred produce genetically identical progeny. A pure breeding tall pea plant is ◊homozygous for the allele controlling height.

putrefaction decomposition of organic matter by microorganisms.

pyramid of biomass in ecology, the declining biomass measured at each trophic level of a ◊food chain. In almost all ecosystems, the producers have the highest total biomass, while the primary and secondary consumers have a greatly reduced total. When these data are represented graphically a pyramid of biomass results. The reason for the reduction is that only around 10% of the energy in one trophic level can pass on to the next, the rest being lost as heat or indigestible matter. For instance, 100 kg of grass will only support 10 kg of herbivore; in turn, the herbivores will only support 1 kg of carnivore.

pyramid of numbers in ecology, the declining numbers of individual organisms that may be counted at each trophic level of a food chain. See ◊pyramid of biomass.

Q

quadrat in ecology, a square structure used to study the distribution of plants in a particular place, for instance a field, rocky shore, or mountainside. The size varies, but is usually 0.5 or 1 metre square, small enough to be carried easily. The quadrat is placed on the ground and the abundance of species estimated. By making such measurements a reliable understanding of species distribution is obtained.

R

radicle the part of a plant embryo that develops into the primary root. Usually it emerges from the seed before the embryonic shoot, or ◊plumule, its tip protected by a root cap, or calyptra, as it pushes through the soil.

radius one of the two bones (the other is the ulna) in the lower forearm of tetrapod (four-limbed) vertebrates.

receptacle the enlarged end of a flower stalk to which the floral parts are attached. Normally the receptacle is rounded, but in some plants it is flattened or cup-shaped.

receptor any cell capable of detecting stimuli. Receptors form part of the nervous system and are used by the body to gather information about the internal or external environment. There are several types, classified according to function. Some respond to light, some to mechanical force, and some to heat. They are essential for ◊homeostasis.

recessive in genetics, term that describes an ◊allele that will show in the ◊phenotype only if its partner allele on the paired chromosome is similarly recessive. Such an allele will not show if its partner is ◊dominant, that is if the organism is ◊heterozygous for a particular characteristic. Alleles for blue eyes in humans, and for shortness in pea plants are recessive. Most mutant alleles are recessive and therefore are only rarely expressed (see ◊haemophilia and ◊sickle-cell disease).

recombination in genetics, any process that recombines, or 'shuffles', the genetic material, thus increasing genetic variation in the offspring. The two main processes of recombination both occur during meiosis (reduction division). One is ◊crossing over, in which chromosome pairs exchange segments; the other is the random reassortment of chromosomes that occurs when each gamete (sperm or egg) receives only one of each chromosome pair.

rectum lowest part of the gut, which stores faeces prior to egestion (defecation).

recycling processing of industrial and household waste (such as paper, glass, and some metals) so that it can be reused, thus saving expenditure on scarce raw materials, slowing down the depletion of non-renewable resources, and helping to reduce pollution.

red blood cell or ***erythrocyte*** the most common type of blood cell, responsible for transporting oxygen around the body. It contains the red protein haemoglobin, which combines with oxygen from the lungs to form oxyhaemoglobin. When transported to the tissues, these cells are able to release the oxygen because the oxyhaemoglobin splits into its original constituents. Mammalian erythrocytes are disclike with a depression in the centre and no nucleus; they are manufactured in the bone marrow and, in humans, last for only four months before being destroyed in the liver and spleen. Those of other vertebrates are oval and nucleated.

reducing sugar sugar that produces a positive result (change in colour from blue to orange) when tested with Benedict's reagent (see ◊food test). All monosaccharides and the disaccharides maltose and lactose are reducing sugars.

reflex in animals, a very rapid automatic response to a particular stimulus. It is controlled by the ◊nervous system. A reflex involves only a few neurons, unlike the slower but more complex responses produced by the many processing neurons of the brain.

A ***simple reflex*** is entirely automatic and involves no learning. Examples of such reflexes include the sudden withdrawal of a hand in response to a painful stimulus, or the jerking of a leg when its knee cap is tapped. Sensory cells (receptors) in the knee send signals to the spinal cord along a sensory neuron. Within the spine a ***reflex arc*** switches the signals straight back to the muscles of the leg (effectors) via an intermediate neuron and then a motor neuron; contraction of the leg muscle occurs, and the leg kicks upwards. Only three neurons are involved, and the brain is only aware of the response after it has taken place. Such reflex arcs are particularly common in lower animals, and

reflex

the knee-jerk reflex arc

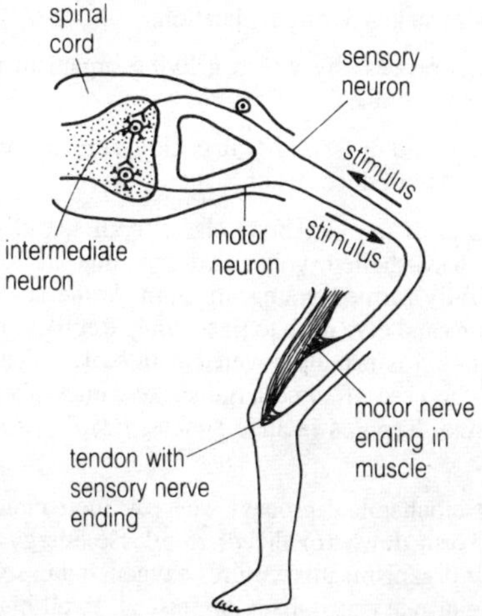

have a high survival value, enabling organisms to take rapid action in order to avoid potential danger. In higher animals (those with a well-developed ◊central nervous system, the simple reflex may be modified by the involvement of the brain—for instance, humans can override the automatic reflex to withdraw a hand from a source of pain.

A ***conditioned reflex*** involves the modification of a reflex action in response to experience (learning). A stimulus that produces a simple reflex response becomes linked with another, possibly unrelated, stimulus. For example, a dog may salivate (a reflex action) when it sees its owner remove a tin-opener from a drawer because it has learned to associate that stimulus with the stimulus of being fed.

relay neuron another name for an ◊intermediate neuron.

rennin or ***chymase*** enzyme found in the gastric juice of young mammals, used in the digestion of milk.

replication the production of copies of the genetic material, DNA; it occurs during cell division (◊mitosis and ◊meiosis). Most mutations are caused by mistakes during replication.

reproduction process by which a living organism produces other organisms similar to itself.

reproduction rate or ***fecundity*** in ecology, the rate at which a population or species reproduces itself.

reptile member of the vertebrate class Reptilia. Unlike amphibians, reptiles have hard-shelled, yolk-filled eggs that are laid on land and from which fully-formed young are born. Some snakes and lizards retain their eggs and give birth to live young. Reptiles are ◊poikilothermic and their skin is usually covered with scales. The metabolism is slow and, in the case of some large snakes, intervals between meals may be months. Reptiles include snakes, lizards, crocodiles, turtles and tortoises.

respiration biochemical process whereby food molecules are progressively broken down (oxidized) to release energy in the form of ◊ATP. In most organisms this requires oxygen, but in some bacteria the oxidant is the nitrate or sulphate ion instead. In all higher organisms, respiration occurs in the ◊mitochondria. Respiration is also used to mean breathing, although this is more accurately described as a form of ◊gas exchange.

respiratory surface area used by an organism for ◊gas exchange—for example, the lungs, gills or, in plants, the leaf interior. The gases oxygen and carbon dioxide are both usually involved in respiration and photosynthesis. Although organisms have evolved different types of respiratory surface according to need, there are certain features in common. These include thinness and moistness, so that the gas can dissolve in a membrane and then diffuse into the body of the organism. In many animals the gas is then transported away from the surface and towards interior cells by the blood system.

respirometer laboratory device for measuring an organism's rate of oxygen uptake.

response any change in an organism occurring as a result of a stimulus. There are many different types of response, some involving the entire organism, others only groups of cells or tissues. Examples include the muscular contractions in an animal, the movement of leaves towards the light, and the onset of hibernation by small mammals at the start of winter. See also ◊behaviour.

retina the inner layer at the back of the ◊eye, which contains light-sensitive cells and neurons (nerve cells). Light falling on the retina produces chemical changes in the light-sensitive cells causing them to send electrical signals along the neuron fibres via the optic nerve to the brain.

The light-sensitive cells are of two types, called rod cells and cone cells after the shape of their outer projections. The ***rod cells***, about 120 million in each eye, are distributed throughout the retina. They are sensitive to low levels of light, but do not provide detailed or sharp images, nor can they detect colour. The ***cone cells***, about 6 million in number, are mostly concentrated in a central region of the retina called the ***fovea***, and provide both detailed and colour vision. The cones of the human eye contain three visual pigments, each of which responds to a different primary colour (red, green, or blue). The brain can interpret the varying signal levels from the three types of cone as any of the different colours of the visible spectrum.

rhesus factor in humans, a protein on the surface of red blood cells that is involved in the rhesus blood group system. Most individuals possess the main rhesus factor (Rh+), but those without this factor (Rh–) produce ◊antibodies if they come into contact with it. The name comes from rhesus monkeys, in whose blood rhesus factors were first found.

If an Rh– mother carries an Rh+ fetus, she may produce antibodies if fetal blood crosses the ◊placenta. This is not normally a problem with the first infant because antibodies are only produced slowly. However, the antibodies continue to build up after birth, and a second Rh+ child may be attacked by antibodies passing from mother to fetus, causing the child to contract anaemia, heart failure, or brain damage. In such cases, the blood of the infant has to be changed for Rh– blood.

rhizome horizontal underground plant stem. It is a ◊perennating organ in some species, where it is generally thick and fleshy, while in other species it is mainly a means of ◊vegetative reproduction, and is therefore long and slender, with buds all along it that send up new plants. The potato is a rhizome that has two distinct parts, the tuber being the swollen end of a long, cordlike rhizome.

rhythm method method of natural contraception that works by avoiding intercourse when the woman is producing ova (ovulating). The time of ovulation can be worked out by the calendar (counting days from the last period), by temperature changes, or by inspection of the cervical mucus. All these methods are unreliable because it is possible for ovulation to occur at any stage of the menstrual cycle.

rib long, usually curved bone that extends laterally from the ◊spine. Most fishes and many reptiles have ribs along most of the spine, but in mammals they are found only in the chest area. In humans, there are 12 pairs of ribs. The ribs protect the lungs and heart, and allow the chest to expand and contract easily.

riboflavin or ***vitamin*** $\boldsymbol{B_2}$ ◊vitamin of the B complex whose absence in the diet causes stunted growth.

ribonucleic acid full name of ◊RNA.

ribosome the protein-making machinery of the cell. Ribosomes are located on the endoplasmic reticulum (ER) of eukaryotic cells. They receive ◊RNA (copied from the ◊DNA in the nucleus) and ◊amino acids, and 'translate' the RNA by using its chemically coded instructions to link the amino acids in a specific order, to make a strand of a particular protein.

rickets defective growth of bone in children because of inadequate calcium deposits. The bones do not harden and are bent out of shape. The condition is usually caused by a lack of calciferol (vitamin D) in the diet and insufficient exposure to sunlight.

RNA (abbreviation for ***ribonucleic acid***) nucleic acid involved in the process of translating ◊DNA, the genetic material into proteins. It is usually single-stranded, unlike the double-stranded DNA, and consists of a large number of nucleotides strung together, each of which comprises of the sugar ribose, a phosphate group, and one of four bases (uracil, cytosine, adenine, or guanine). RNA is copied from DNA by the formation of ◊base pairs, with uracil taking the place of thymine.

rod cell a type of light-sensitive cell found in the ◊retina of the eye. Rods are highly sensitive and provide only black-and-white vision. They are used when lighting conditions are poor and are the only type of visual cell found in animals active at night.

root the part of a plant that is usually underground, and whose primary functions are anchorage and the absorption of water and dissolved mineral salts. Roots usually grow downwards and towards water (that is, they are positively geotropic and hydrotropic; see ◊tropism).

The absorptive area of roots is greatly increased by the numerous, slender ◊root hairs formed near the tips. A root cap, or calyptra, protects the tip of the root from abrasion as it grows through the soil.

Symbiotic associations occur between the roots of certain plants, such as clover, and various bacteria that fix nitrogen from the air (see ◊nitrogen fixation).

root cap or ***calyptra*** cap at the tip of a growing root. It gives protection to the zone of actively dividing cells as the root pushes through the soil.

root hair tiny hairlike outgrowth on the surface cells of plant roots that greatly increases the area available for the absorption of water and other materials. New root hairs are continually being formed near the root tip, one of the places where plants show the most active growth. The hairs are extremely delicate.

root pressure water pressure developed within the roots of plants because of their active uptake of minerals from the soil. It and the ◊transpiration stream account for the upward movement of water through the stem.

roughage or ***dietary fibre*** part of the diet that is indigestible, consisting mostly of cellulose from plant cell walls. It adds bulk to the gut contents, assisting the process of ◊peristalsis, the muscular contractions forcing the food along the intestine. A high roughage content is

root

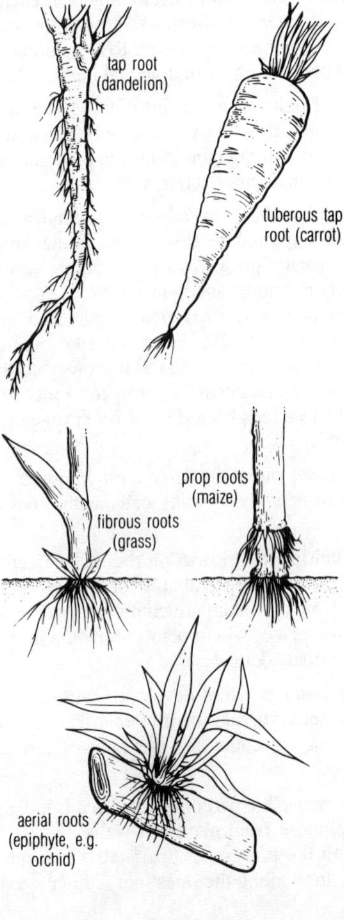
tap root
(dandelion)
tuberous tap
root (carrot)
prop roots
(maize)
fibrous roots
(grass)
aerial roots
(epiphyte, e.g.
orchid)

believed to have several beneficial effects, including reduced cancer risks.

rubber another name for a ◊condom.

runner aerial stem that produces new plants. It is a means of ◊vegetative reproduction.

S

saccharide another name for a ◊sugar molecule.

saliva in vertebrates, a secretion from the salivary glands that aids the swallowing and digestion of food in the mouth. In mammals, it contains the enzyme amylase, which converts starch to sugar.

salmonella group of bacteria living in the intestines of animals including humans. Some cause diseases such as typhoid and paratyphoid fevers; others cause salmonella food poisoning, which is characterized by stomach pains, vomiting, diarrhoea, and headache. Most cases of salmonella food poisoning are caused by contaminated animal products, especially poultry meat.

saprotroph (formerly ***saprophyte***) an organism that feeds on the excrement, or the dead bodies or tissues of others. They include most fungi (the rest being parasites); many bacteria and protozoa; animals such as dung beetles and vultures; and a few unusual plants, including several orchids. Saprotrophs cannot make food for themselves, so they are a type of ◊heterotroph. They are useful scavengers, and in sewage farms and refuse dumps break down organic matter into nutrients easily assimilable by green plants.

scapula or ***shoulder blade*** large bone forming part of the pectoral girdle, assisting in the articulation of the arm with the chest region. Its flattened shape allows a large region for the attachment of muscles.

scent gland a gland that opens onto the outer surface of animals, producing odorous compounds that are used for communicating between members of the same species (◊pheromones), or for discouraging predators.

sclerenchyma a plant tissue whose function is to strengthen and support, composed of thick-walled cells that are heavily lignified (toughened). On maturity the cell inside dies, and only the cell walls remain.

Sclerenchyma may be made up of one or two types of cells: ***sclereids***, occurring singly or in small clusters, are often found in the hard shells of fruits and in seed coats, bark, and the stem cortex; ***fibres***, frequently grouped in bundles, are elongated cells, often with pointed ends, associated with the vascular tissue (◊xylem and ◊phloem) of the plant.

scurvy disease caused by deficiency of ascorbic acid (vitamin C), which is contained in fresh vegetables and fruit. The signs are weakness and aching joints and muscles, progressing to bleeding of the gums and then other organs, and drying-up of the skin and hair. Treatment is by giving the vitamin.

secondary growth or ***secondary thickening*** the increase in diameter of the roots and stems of certain plants (notably shrubs and trees) that results from the production of new cells by the ◊cambium (lateral meristem). It provides the plant with additional mechanical support and new conducting cells, the secondary ◊xylem and ◊phloem. Secondary growth is generally confined to ◊gymnosperms and, among the ◊angiosperms, to the dicotyledons. With just a few exceptions, the monocotyledons (grasses, lilies) exhibit only primary growth, resulting from cell division at the apical ◊meristems.

secondary sexual characteristic external feature of an organism that is characteristic of its gender (male or female), but not the reproductive organs themselves. Secondary sexual characteristics include facial hair in men and breasts in women, combs in cockerels, brightly coloured plumage in many male birds, and manes in male lions. In many cases, they are involved in displays and contests for mates and have evolved by ◊sexual selection. Their development is stimulated by sex hormones.

secretion any substance (normally a fluid) produced by a cell or specialized gland, for example, sweat, saliva, enzymes, and hormones. The process whereby the substance is discharged from the cell is also known as secretion.

sedative drug that lessens nervousness, excitement, or irritation. Sedatives will induce sleep in larger doses.

seed the reproductive structure of higher plants (◊angiosperms and ◊gymnosperms). It develops from a fertilized ovule and consists of an embryo and a food store, surrounded and protected by an outer seed coat, called the testa. The food store is contained either in a specialized nutritive tissue, the ◊endosperm, or in the ◊cotyledons of the embryo itself. In angiosperms the seed is enclosed within a ◊fruit, whereas in gymnosperms it is usually naked and unprotected, once shed from the female cone. Following ◊germination the seed develops into a new plant.

Seeds may be dispersed from the parent plant in a number of different ways. Agents of dispersal include animals, as with ◊burs and fleshy edible fruits, and wind, where the seed or fruit may be winged or plumed. Water can disperse seeds or fruits that float, and various mechanical devices may eject seeds from the fruit, as in some pods or ◊leguminous plants.

seed plant or ***spermatophyte*** any seed-bearing plant. The seed plants are subdivided into two classes, the ◊angiosperms, or flowering plants, and the ◊gymnosperms, principally the conifers and cycads. Together, they comprise the major types of vegetation found on land.

Angiosperms are the largest, most advanced, and most successful group of plants at the present time, occupying a highly diverse range of habitats. There are estimated to be about 250,000 different species. Gymnosperms differ from angiosperms in their ovules which are borne unprotected (not within an ◊ovary) on the scales of their cones. The arrangement of the reproductive organs, and their more simplified internal tissue structure, also distinguishes them from the flowering plants. In contrast to the gymnosperms, the ovules of angiosperms are enclosed within an ovary and many species have developed highly specialized reproductive structures associated with ◊pollination by insects, birds, or bats.

semen fluid containing ◊sperm from the testes and secretions from various sex glands, such as the prostate gland, that is produced by male animals during copulation. The secretions serve to nourish and activate the sperm cells, and prevent their clumping together.

semicircular canal one of three looped tubes that form part of the labyrinth in the inner ◊ear. They are filled with fluid and detect changes in the position of the head, contributing to the sense of balance.

sense organ any organ that an animal uses to gain information about its surroundings. All sense organs have specialized receptors (such as light receptors in an eye) and some means of translating their response into a nerve impulse that travels along sensory neurons to the brain. The main human sense organs are the ◊eye, which detects light and colour (different wavelengths of light); the ◊ear, which detects sound (vibrations of the air) and gravity; the ◊nasal cavity, which detects some of the chemical molecules in the air; and the ◊tongue, which detects some of the chemicals in food, giving a sense of taste. There are also many small sense organs in the skin, including pain sensors, temperature sensors, and pressure sensors, contributing to our sense of touch.

Some animals can detect small electric discharges, underwater vibrations, minute vibrations of the ground, or sounds that are below (infrasound) or above (ultrasound) the range of hearing in humans. Sensitivity to light varies greatly. Most mammals cannot distinguish different colours, whereas many insects can see light in the ultraviolet range, and snakes can form images of infrared radiation (radiant heat).

sensitivity the ability of an organism, or part of an organism, to detect changes in the environment. Although all living things are capable of some sensitivity, evolution has led to the formation of highly complex mechanisms for detecting light, sound, chemicals, and other stimuli. It is essential to an animal's survival that it can process this type of information and make an appropriate response.

sepal part of a flower, usually green, that surrounds and protects the flower in bud. The sepals are derived from modified leaves, and collectively known as the ◊calyx.

In some plants, such as the marsh marigold, where true ◊petals are absent, the sepals are brightly coloured and petal-like, taking over the role of attracting insect pollinators to the flower.

serum a clear fluid that remains after blood has clotted. It is blood plasma with the anticoagulant proteins removed, and contains

sexual reproduction in humans

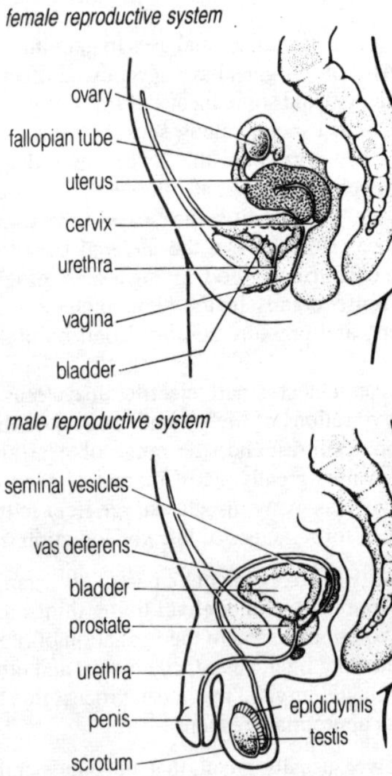

◊antibodies and other proteins, as well as the fats and sugars of the blood. It can be produced synthetically, and is used to protect against disease.

sex chromosome one of the pair of chromosomes that determines the sex of an organism.

Maleness and femaleness in animals usually depends on which of the sex chromosomes have been inherited. The two types of chromosome, known as the X and Y chromosomes, are paired in the adult organism. Human beings are male when they possess XY and female when they possess XX. An ovum (egg) or a sperm cell is haploid and can only contain one sex chromosome so all ova are X. Sperm however may be X or Y; the zygote resulting from fusion of human gametes will be male if the sperm cell contained a Y chromosome.

sex linkage in genetics, the tendency for certain characteristics to occur exclusively, or predominantly, in one sex only. Human examples include red–green colour blindness and haemophilia, both found predominantly in males. In both cases, these characteristics are ◊recessive and are determined by genes on the ◊X chromosome.

Since females possess two X chromosomes, any such recessive ◊allele on one of them is likely to be masked by the corresponding allele on the other. In males (who have only one X chromosome paired with a largely inert ◊Y chromosome) any gene on the X chromosome will automatically be expressed. Colour blindness and haemophilia can appear in females, but only if they are ◊homozygous for these traits, due to inbreeding, for example.

sexual reproduction a reproductive process in organisms that requires the union, or ◊fertilization, of gametes (such as eggs and sperm). These are usually produced by two different individuals, although self-fertilization occurs in a few ◊hermaphrodites such as tapeworms. Most organisms other than bacteria and cyanobacteria show some sort of sexual process. Except in some lower organisms, the gametes are of two distinct types called ova (eggs) and sperm. The organisms producing the ova are called females, and those producing the sperm, males. The fusion of a male and female gamete produces a *zygote*, from which a new individual develops. The alternatives to sexual reproduction are binary fission, budding, vegetative reproduction, parthenogenesis, and spore formation.

sexual selection a process similar to ◊natural selection but relating exclusively to success in finding a mate for the purpose of sexual reproduction and producing offspring. Sexual selection occurs when one sex (usually but not always the female) invests more effort in producing young than the other. Members of the other sex compete for

sexual reproduction in flowerng plants

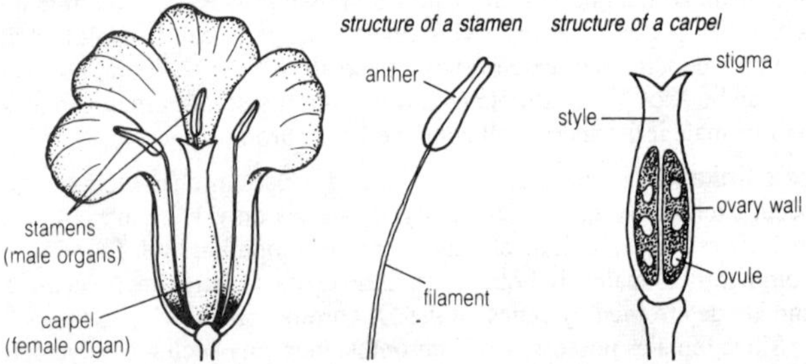

access to this limited resource (usually males compete for the chance to mate with females). Sexual selection often favours features that increase a male's attractiveness to females (such as the pheasant's 'tail') or enable males to fight with one another (such as a deer's antlers). More subtly, it can produce hormonal effects by which the male makes the female unreceptive to other males, causes the abortion of fetuses already conceived, or removes the sperm of males who have already mated with a female.

sheath another name for a ◊condom.

shoot general term for those parts of a ◊vascular plant that grow above ground, comprising a stem bearing leaves, buds, and flowers. The shoot develops from the ◊plumule of the embryo.

short sight or ***myopia*** the inability of the ◊eye to focus on distant objects. Short sight occurs when the lens loses its ability to become flatter, or when the eyeball is too long. As a result, the image is focused on a point in front of a retina. The problem can be corrected by use of a diverging or concave lens.

sight the detection of light by an ◊eye, which can form images of the outside world.

skeleton the rigid or semirigid framework that supports an animal's body, protects its internal organs, and provides anchorage points for its muscles. The skeleton may be composed of bone and cartilage (vertebrates), chitin (arthropods), calcium carbonate (molluscs and some other invertebrates), or silica (many unicellular eukaryotes).

It may be internal, forming an ◊endoskeleton, or external, forming an ◊exoskeleton. Another type of skeleton, found in invertebrates such as earthworms, is the ***hydrostatic skeleton***. This gains partial rigidity from fluid enclosed within a body cavity. Because the fluid cannot be compressed, contraction of one part of the body results in extension of another part, giving peristaltic motion.

skin in vertebrates, the covering of the body. In mammals, the outer layer (epidermis) is dead and protective, and its cells are constantly being rubbed away and replaced from below. The lower layer (dermis) contains blood vessels, nerves, hair roots, and sweat and sebaceous glands, and is supported by a network of fibrous and elastic cells.

skull in vertebrates, the collection of flat and irregularly shaped bones (or cartilage) that enclose the brain and the organs of sight, hearing, and smell, and provide support for the jaws. In mammals, the skull consists of 22 bones joined by sutures (immovable joints). The floor of the skull is pierced by a large hole for the spinal cord and a number of smaller apertures through which other nerves and blood vessels pass.

sleep a state of reduced awareness and activity that occurs at regular intervals in most mammals and birds, though there is considerable variation in the amount of time spent sleeping. Sleep differs from hibernation in that it occurs daily rather than seasonally, and involves less drastic reductions in metabolism. The function of sleep is unclear. People deprived of sleep become irritable, uncoordinated, and forgetful.

small intestine the length of gut between the stomach and the large intestine, consisting of the duodenum and the ileum. It is responsible for digesting and absorbing food.

The wall is glandular, producing mucus and enzymes to aid digestion, and muscular so that food can be moved down the gut. Absorption, the passage of small molecules across the wall, is made more efficient by the presence of numerous villi, small finger-like projections that increase the surface area.

smell sense that responds to chemical molecules in the air. It is used to detect food and to communicate with other animals (see ◊pheromone and ◊scent gland). It works by having receptors (◊olfactory receptors) for particular chemical groups, into which the airborne chemicals must fit to trigger a message to the brain. See also ◊nasal cavity.

smoking inhaling the fumes from burning tobacco, usually in the form of cigarettes. The practice can be habit-forming and is dangerous to health, since carbon monoxide and other poisonous materials result from the combustion process. There is a direct link between lung cancer and tobacco smoking; the habit is also linked to respiratory conditions such as bronchitis and emphysema, and to coronary heart diseases. In the West, smoking is now forbidden in many public places because even ***passive smoking***—breathing in fumes from other people's cigarettes—can be harmful. Children whose parents smoke suffer an increased risk of asthma and respiratory infections.

Manufacturers have attempted to filter out harmful substances such as tar and nicotine, and to use milder tobaccos, and governments have carried out extensive antismoking advertising campaigns. In the UK and the USA all cigarette packaging must carry a government health warning, and television advertising of cigarettes is forbidden.

smooth muscle another name for ◊involuntary muscle.

social behaviour behaviour concerned with altering the behaviour of other individuals of the same species. Social behaviour allows animals to live harmoniously in groups by establishing hierarchies of dominance to discourage disabling fighting. It may be aggressive or submissive (for example, cowering and other signals of appeasement), or designed to establish bonds (such as social grooming or preening).

The social behaviour of mammals and birds is generally more complex than that of lower organisms, and involves relationships with individually recognized animals. Thus, courtship displays allow individuals to choose appropriate mates and form the bonds necessary for successful reproduction. In the social systems of bees, wasps, ants, and termites, an individual's status and relationships with others is largely determined by its biological form, as a member of a caste of workers, soldiers, or reproductive.

soil

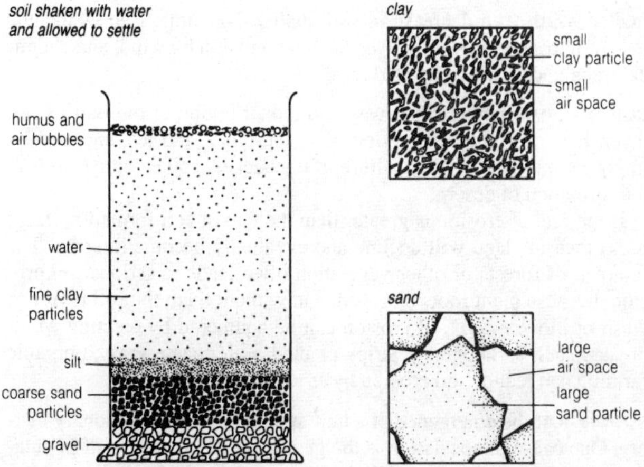

soil loose mixture of finely ground rock and decaying organic material covering the Earth's surface. It provides nutrients for plants, but can lose its fertility if these nutrients are not replenished. The decomposition of dead animals and plants in the upper surface of the soil—the humus layer—continually restores the organic and mineral content. Aeration and mixing is also necessary for a healthy soil, and this is often achieved by the burrowing activities of earthworms.

There are many types, with varying organic and mineral content making some soils more suitable than others for particular types of natural vegetation or for agriculture. Particle size is responsible for one of the major differences between soils. Clay soils have very tiny particles, with only small spaces between each grain, leading to poor drainage qualities. Sandy soils have large particles, but can drain too easily, leading to drying out and, in some cases, desertification. The organic content of soil is widely variable, ranging from zero in some desert soils to almost 100% in peats. Loams are very fertile soils containing a mixture of organic material and particles of different sizes.

soil depletion a decrease in soil quality over time. Causes include loss of nutrients caused by over-farming, erosion by wind, and chemical imbalances caused by acid rain.

soil erosion the wearing away and redistribution of the Earth's soil layer. It is caused by the action of water, wind, and ice, and also by improper methods of ◊agriculture. If unchecked, soil erosion results in the formation of deserts.

If the rate of erosion is greater than the rate of soil formation (from rock) then the land will decline and eventually become infertile. The removal of forests or other vegetation often leads to serious soil erosion, because plant roots bind soil, and without them the soil is free to wash or blow away. Soil erosion can be countered by erecting windbreaks, such as hedges or strips planted with coarse grass; ◊organic farming can reduce soil erosion by as much as 75%.

speciation the emergence of a new species during evolutionary history. One cause of speciation is the geographical separation of populations of the parent species, followed by reproductive isolation, so that they no longer produce viable offspring unless they interbreed.

species the lowest level of classification; a distinguishable group of organisms that resemble each other or consist of a few distinctive types (varieties), and that can all interbreed to produce fertile offspring.

Related species are grouped together in a genus. Within a species there are usually two or more separate populations, which may in time become distinctive enough to be designated subspecies or varieties, and could eventually give rise to new species through ◊speciation. Around 1.4 million species have been identified so far, of which 750,000 are insects, 250,000 are plants, and 41,000 are vertebrates.

All living human beings belong to the same species because they can all interbreed, even though they may differ considerably in such features as skin colour, height, head shape, and so on. Other examples of species are lions, Douglas firs, cabbage white butterflies, and sperm whales.

sperm

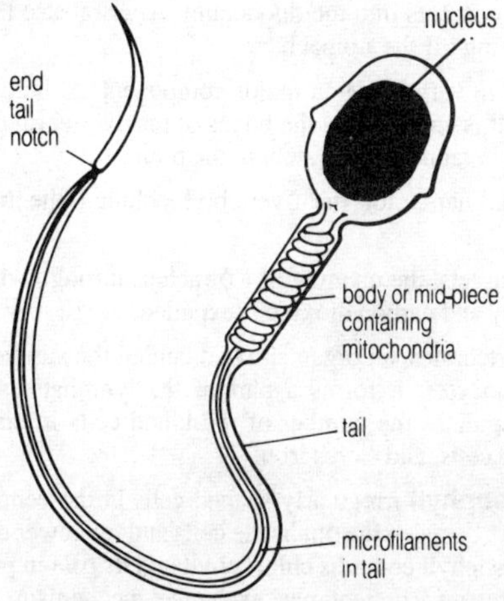

sperm or ***spermatozoon*** the male ◊gamete of animals. Each sperm cell has a head capsule containing a nucleus, a middle portion containing ◊mitochondria (which provide energy), and a long tail (flagellum). In most animals, the sperm are motile, and are propelled by the flagellum. The sperm cells of animals that carry out internal ◊fertilization are usually released, with secretions from various sex glands, in the form of a fluid called ◊semen. Sperm cells are produced in the testes (see ◊testis).

spermicide any cream, jelly, pessary, or other preparation that kills the sperm cells in semen. Spermicides are used for contraceptive purposes, usually in combination with a ◊condom or ◊diaphragm. Spermicide used alone is only 75% effective in preventing pregnancy.

sphincter ring of muscle found at various points in the alimentary canal, which contracts and relaxes to control the movement of food.

The ***pyloric sphincter***, at the base of the stomach, controls the release of the gastric contents into the duodenum. After release the sphincter contracts, closing off the stomach.

spinal cord in vertebrates, a major component of the ◊central nervous system. It is enclosed by the bones of the ◊vertebral column, and links the peripheral nervous system to the brain.

spine another name for the ◊vertebral column, the backbone of vertebrates.

spiracle in insects, the opening of a ◊trachea, through which oxygen enters the body and carbon dioxide is expelled.

spleen in vertebrates, an organ situated behind the stomach that produces ◊lymphocytes; it forms a part of the lymphatic system. The spleen also regulates the number of red blood cells in circulation by destroying old cells, and stores iron.

spongy mesophyll irregularly shaped cells in the centre of a leaf, forming a layer between the ◊palisade cells and the lower epidermis.

Spongy mesophyll contains chlorophyll and its role in photosynthesis is aided by an efficient gas exchange mechanism. Air spaces between the cells are continuous with the stomata (see ◊stoma), small pores found on the undersurface of the leaf. Gases can circulate freely, carbon dioxide entering the cells if needed, and oxygen passing out of the leaf as an excretory product.

sporangium a structure in which ◊spores are produced.

spore a small reproductive or resting body, usually consisting of just one cell. Unlike a ◊gamete, it does not need to fuse with another cell in order to develop into a new organism. Spores are produced by the lower plants, most fungi, some bacteria, and certain protozoa. They are generally light and easily dispersed by wind movements.

Plant spores are haploid and are produced by the sporophyte, following ◊meiosis; see ◊alternation of generations.

stamen in flowers, the male reproductive organ. The stamens are collectively referred to as the ◊androecium. A typical stamen consists of a ***filament***, with an ***anther***, the pollen-bearing organ, at its apex.

stamen

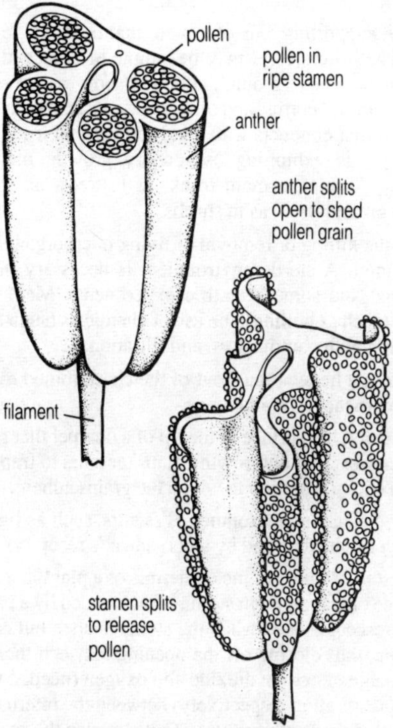

starch polysaccharide made up of monosaccharide glucose units. It is produced by plants as a carbohydrate food reserve, and is stored in the form of ***starch grains*** in the cells of roots, tubers, seeds, and fruit. Starch is therefore a important source of energy for many primary consumers (herbivores and omnivores). For humans, the main dietary sources of starch are cereals, legumes (beans and peas), potatoes, and root vegetables.

stem the main supporting axis of a plant that bears the leaves, buds, and reproductive structures; it may be simple or branched. The plant stem usually grows above ground, although some grow underground, including ◊rhizomes, ◊corms, and ◊tubers. Stems contain a continuous vascular system that conducts water and food to and from all parts of the plant. In plants exhibiting ◊secondary growth, the stem may become woody, forming a main trunk, as in trees, or a number of branches from ground level, as in shrubs.

sterilization the killing or removal of living microorganisms such as bacteria and fungi. A sterile environment is necessary in medicine, food processing, and some scientific experiments. Methods include heat treatment (such as boiling), the use of chemicals (such as disinfectants), irradiation with gamma rays, and filtration.

sternum large flat bone at the front of the chest, joined to the ribs. It gives protection to the heart and lungs.

stigma in a flower, the surface at the tip of a ◊carpel that receives the ◊pollen. It often has short outgrowths, flaps, or hairs to trap pollen and may produce a sticky secretion to which the grains adhere.

stimulus any change in environmental factors, such as light, heat, or pressure, which can be detected by an organism's receptors.

stoma plural ***stomata*** pore in the epidermis of a plant, particularly on the undersurface of leaves. Each stoma is surrounded by a pair of guard cells that are crescent shaped when the stoma is open but can collapse to an oval shape, thus closing off the opening between them. Stomata allow the exchange of carbon dioxide and oxygen (needed for ◊photosynthesis and ◊respiration respectively) between the internal tissues of the plant and the outside atmosphere. They are also the main route by which water is lost from the plant by ◊transpiration, and they can be closed to conserve water, the movements being controlled by changes in turgidity of the guard cells.

stomach the first cavity in the gut. In mammals it is a bag of muscle situated just below the diaphragm. Food enters it from the oesophagus,

stoma

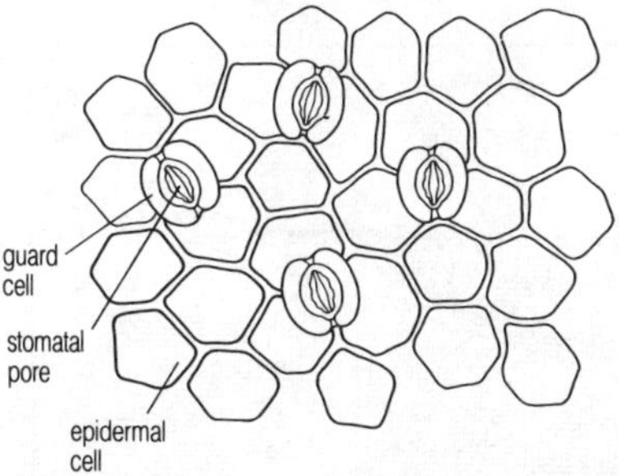

is digested by the acid and ◊enzymes secreted by the stomach lining, and then passes into the duodenum. Some plant-eating mammals, such as cows and sheep, have multichambered stomachs that harbour bacteria in one of the chambers to assist in the digestion of ◊cellulose.

striped muscle or ***striated muscle*** muscle tissue responsible for most types of movement. It is attached to the skeleton, and contracts and relaxes under voluntary nervous control.

Under the microscope it has a distinctive striped appearance, resulting from the coordinated arrangement of protein molecules within the fibres. Unlike ◊involuntary muscle, fast contraction is possible.

style in flowers, the part of the ◊carpel bearing the ◊stigma at its tip. In some flowers it is very short or completely lacking, while in others it may be long and slender, positioning the stigma in the most effective place to receive the pollen.

substrate compound or mixture of compounds that is acted upon by an enzyme. The term also refers to a substance such as ◊agar that provides the nutrients for the metabolism of microorganisms.

succession in ecology, series of changes that occur in the structure and composition of the vegetation in a given area from the time it is first colonized by plants (***primary succession***), or after it has been disturbed by fire, flood, or clearing (***secondary succession***).

succession

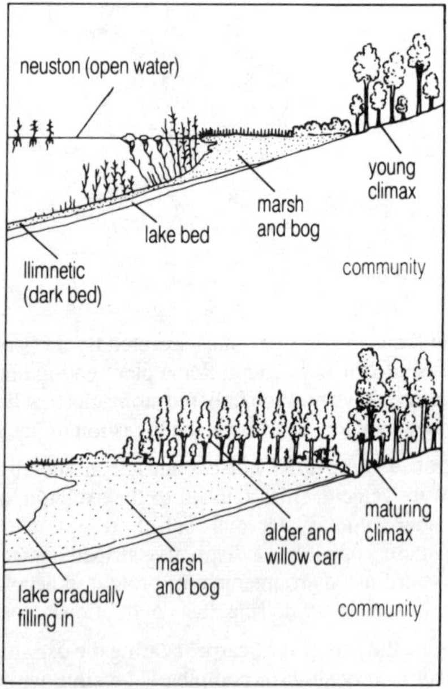

If allowed to proceed undisturbed, succession leads naturally to a stable ◊climax community (for example, oak forest or savannah) that is determined by the climate and soil characteristics of the area.

succulent plant a thick, fleshy plant that stores water in its tissues—for example, cacti. Succulents live either in areas where water is very scarce, such as deserts, or in places where it is not easily obtainable because of the high concentrations of salts in the soil, as in salt marshes. See also ◊xerophyte.

suckering in plants, form of ◊vegetative reproduction by which new shoots (suckers) arise from an existing root system rather than from seed. Plants that produce suckers include elm, dandelion, and members of the rose family.

sucrose or ***cane sugar*** or ***beet sugar*** ($C_{12}H_{22}O_{10}$) ◊disaccharide sugar made up of two monosaccharides glucose and fructose, obtained commercially from cane or from sugar beet. It is a non-reducing sugar (see ◊food test).

sugar soluble ◊carbohydrate, either a monosaccharide or disaccharide, usually with a sweet taste. The term is popularly used to refer to sucrose (cane or beet sugar) only.

Sugars may be described as reducing (all monosaccharides and the disaccharides maltose and lactose) or non-reducing (all other disaccharides) depending on whether they produce a positive result (change in colour) when tested with Benedict's solution (see ◊food test).

support the mechanism by which an organism holds itself and maintains its shape in relation to gravity.

Plants support themselves by keeping their cells turgid (see ◊turgor) and, in the case of higher plants, by the rigidity of their vascular bundles (xylem and phloem). Large plants such as trees and shrubs undergo ◊secondary growth and strengthen their xylem vessels with lignin to form tough wood.

Animals tend to support themselves by means of a hard skeleton, which may be either internal (◊endoskeleton) or external (◊exoskeleton). It is important that such structures should not impede the motion of the animal, therefore most skeletons have movable joints. Exoskeletons are usually shed periodically to allow growth, while endoskeletons grow with the developing organism. Certain soft invertebrates such as earthworms make use of the pressure exerted on their body walls by their internal fluids in order to maintain a semirigid structure. Many aquatic organisms, both plant and animal, rely on the buoyancy provided by water for their support.

suprarenal gland alternative name for the ◊adrenal gland.

surface-area-to-volume ratio the ratio of an animal's surface area (the area covered by its skin) to its total volume. This is high for small animals, but low for large animals such as elephants.

The ratio is important for homeothermic (warm-blooded) animals because the amount of heat lost by the body is proportional to its surface area, whereas the amount generated is proportional to its volume. Very small birds and mammals, such as hummingbirds and shrews, lose a lot of heat and need a high intake of food to maintain their body temperature. Elephants, on the other hand, are in danger of overheating, which is why they have no fur.

suspensory ligament in the ◊eye, a ring of fibre supporting the lens. The ligaments themselves attach to the ciliary muscles, the circle of muscle mainly responsible for changing the shape of the lens during ◊accommodation. If the ligaments are put under tension, the lens becomes flatter, and therefore able to focus on objects in the far distance.

sweat gland an ◊exocrine gland within the skin of mammals that produces surface perspiration. In primates, sweat glands are distributed over the whole body, but in most other mammals they are more localized; for example, in cats and dogs, they are restricted to the feet and around the face.

symbiosis any close relationship between two organisms of different species, and one where both partners benefit from the association. A well-known example is the pollination relationship between insects and flowers, where the insects feed on nectar and carry pollen from one flower to another. This is sometimes known as ◊mutualism. Symbiosis in a broader sense includes ◊commensalism and parasitism (see ◊parasite).

synapse the junction between two ◊neurons (nerve cells), or between a neuron and a muscle (a neuromuscular junction), across which a nerve impulse is transmitted. The two cells involved are not in direct contact but separated by a narrow gap called the ***synaptic cleft***. The threadlike extension, or ◊axon, of the transmitting neuron has a slightly swollen terminal point, the ***synaptic knob***. This forms one half of the synaptic junction and houses membrane-bound vesicles, which contain a chemical ◊neurotransmitter. When nerve impulses reach the knob, the vesicles release the transmitter and this flows across the gap and

synapse

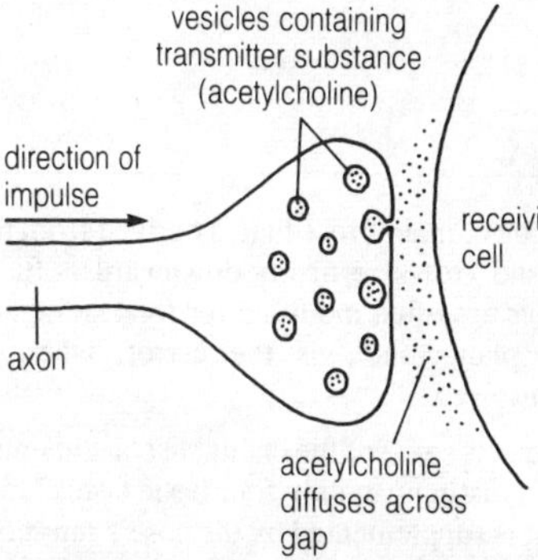

binds itself to special receptors on the receiving cell's membrane. If the receiving cell is a neuron, the other half of the synaptic junction will be one or more extensions called ◊dendrites; these will be stimulated by the neurotransmitter to set up an impulse, which will then be conducted along the length of the neuron and on to its own axons. If the receiving cell is a muscle cell, it will be stimulated by the neurotransmitter to contract.

synovial fluid a viscous yellow fluid that bathes movable joints between the bones of vertebrates. It nourishes and lubricates the ◊cartilage at the end of each bone.

system a collection of organs and tissues that work together to perform one or more coordinated functions. The blood system involves many structures—the heart, the blood, the blood vessels—all working together to transport materials around the body. Similarly, the nervous system is able to respond to a stimulus because the muscles, brain, and nerves are working as a coordinated system. Other systems include the digestive and urinary systems.

T

taproot single, robust, main ◊root that is derived from the embryonic root, or ◊radicle, and grows vertically downwards, often to considerable depth. Taproots are often modified for food storage and are common in biennial plants such as the carrot, where they act as ◊perennating organs.

taste sense that detects some of the chemical constituents of food. The human ◊tongue can distinguish only four basic tastes (sweet, sour, bitter, and salty) but it is supplemented by the nose's sense of smell. What we refer to as taste is really a composite sense made up of both taste and smell.

taxis (plural ***taxes***) or ***tactic movement*** the movement of a single cell, such as a bacterium, protozoan, single-celled alga, or gamete, in response to an external stimulus. A movement directed towards the stimulus is described as positive taxis; a movement away from it as negative taxis. The alga *Chlamydomonas*, for example, demonstrates positive ***phototaxis*** by swimming towards a light source to increase the rate of photosynthesis. ***Chemotaxis*** is a response to a chemical stimulus, as seen in many bacteria that move towards higher concentrations of nutrients.

taxonomy another name for the ◊classification of living organisms.

temperature regulation the ability of an organism to control its body temperature.

Although some plants have evolved ways of resisting extremes of temperature, sophisticated mechanisms for maintaining the correct temperature are found in multicellular animals. Such mechanisms may be behavioural, as when a lizard moves into the shade in order to cool down. Mammals and birds have internal control and are known as ***homeotherms***. These animals are insulated with fat, hair, or feathers to conserve heat produced by metabolic activities. Other adaptations allow heat to leave the body when the animal is in danger of overheating, for instance during intense activity. Such mechanisms include sweating, increased flow of blood through the skin (vasodilation), and panting.

tendon or ***sinew*** a cord of tough, fibrous connective tissue that joins muscle to bone in vertebrates. Tendons are largely composed of the protein collagen, and because of their inelasticity are very efficient at transforming muscle power into movement.

tendril slender, threadlike structure that supports a climbing plant by coiling around suitable supports, such as the stems and branches of other plants. It may be a modified stem, leaf, leaflet, flower, leaf stalk, or stipule (a small appendage on either side of the leaf stalk), and may be simple or branched.

territory in behaviour, a fixed area from which an animal or group of animals excludes other members of the same species. Animals may hold territories for many different reasons; for example, to provide a constant food supply, to monopolize potential mates, or to ensure access to refuges or nest sites. The size of a territory depends in part on its function: some nesting and mating territories may be only a few square metres, whereas feeding territories may be as large as hundreds of square kilometres.

testa the outer coat of a seed, formed after fertilization of the ovule. It has a protective function and is usually hard and dry. In some cases the coat is adapted to aid dispersal, for example by being hairy. Humans have found uses for many types of testa, including the fibre of the cotton seed.

test cross or ***back cross*** in genetics, a breeding experiment used to discover the genotype of an individual organism for a particular characteristic. If the test individual is crossed with a double recessive of the same species, its offspring will indicate whether it is homozygous or heterozygous for the characteristic in question. In peas, a tall plant under investigation would be crossed with a double recessive short plant with known genotype tt. The results of the cross will be all tall plants if the test plant is TT. If the individual is in fact Tt then there will be some short plants (genotype tt) among the offspring.

testis (plural ***testes***) the organ that produces ◊sperm in male animals. In vertebrates it is one of a pair of oval structures that are usually internal, but in mammals (other than elephants and marine mammals), the paired testes (or testicles) descend from the body cavity during development, to hang outside the abdomen in a sac called the ***scrotum***.

testosterone in vertebrates, hormone secreted chiefly by the testes. It promotes the development of ◊secondary sexual characteristics in males. In animals with a breeding season, the onset of breeding behaviour is accompanied by a rise in the level of testosterone in the blood.

tetrapod four-legged vertebrate. The tetrapods include mammals, birds, reptiles, and amphibians. Birds are included because they evolved from four-legged ancestors, the forelimbs having become modified to form wings. Even snakes are tetrapods, since their lack of limbs is secondary.

thiamine or ***vitamin*** $\boldsymbol{B_1}$ ◊vitamin of the B complex. Its absence from the diet causes the disease ◊beriberi.

thorax in tetrapod vertebrates, the part of the body containing the heart and lungs, and protected by the rib cage. It is separated from the abdomen by the muscular diaphragm in mammals.

In arthropods, such as insects, the thorax is the middle part of the body, between the head and abdomen.

throat in humans, the passage that leads from the back of the nose and mouth to the ◊trachea and ◊oesophagus. It includes the ◊pharynx and the ◊larynx, the latter being at the top of the trachea. The term is also used to mean the front part of the neck, both in humans and other vertebrates.

thyroid ◊endocrine gland situated in the neck in front of the trachea. It secretes several hormones, among them thyroxin, a hormone that stimulates growth, metabolism, and other functions of the body.

tibia the anterior of the pair of bones found between the ankle and the knee. In humans, the tibia is the shinbone.

tissue a collection of cells that perform a similar function. Thus, nerve and muscle are different kinds of tissue in animals, as are ◊parenchyma and ◊sclerenchyma in plants.

tissue culture process by which cells from a plant or animal are removed from the organism and grown under controlled conditions in a sterile medium containing all the necessary nutrients. Tissue culture can provide information on cell growth and differentiation, and is also used in the propagation of plants. See also ◊meristem.

tongue in tetrapod vertebrates, a muscular organ usually attached to the floor of the mouth. It has a thick root attached to a U-shaped bone, and is covered with a ◊mucous membrane containing nerves and 'taste buds'. It directs food to the teeth and into the throat for chewing and swallowing. In humans, it is crucial for speech; in other animals, for lapping up water and for grooming, among other functions.

tonsils in higher vertebrates, masses of lymphoid tissue situated at the back of the mouth and throat, and on the rear surface of the tongue. The tonsils contain many ◊lymphocytes and are part of the body's defence system against infection.

tooth in vertebrates, one of a set of hard, bonelike structures in the mouth, used for biting and chewing food, and in defence and aggression. In humans, the first set (20 milk teeth) appear from age six months to two and a half years. The permanent ◊dentition replaces these from the sixth year onwards, the wisdom teeth (third molars) sometimes not appearing until the age of 25 or 30. Adults have 32 teeth: two incisors, one canine, two premolars, and three molars on each side of each jaw. Each tooth consists of an enamel coat (hardened calcium deposits), dentine (a thick, bonelike layer), and an inner pulp cavity, housing nerves and blood vessels. Mammalian teeth have roots surrounded by cement, which fuses them into their sockets in the jawbones. The neck of the tooth is covered by the ◊gum, while the enamel-covered crown protrudes above the gum line.

touch sensation produced by specialized nerve endings (receptors) in the skin. Some respond to light pressure, others to heavy pressure. Temperature detection may also contribute to the overall sensation of

tooth

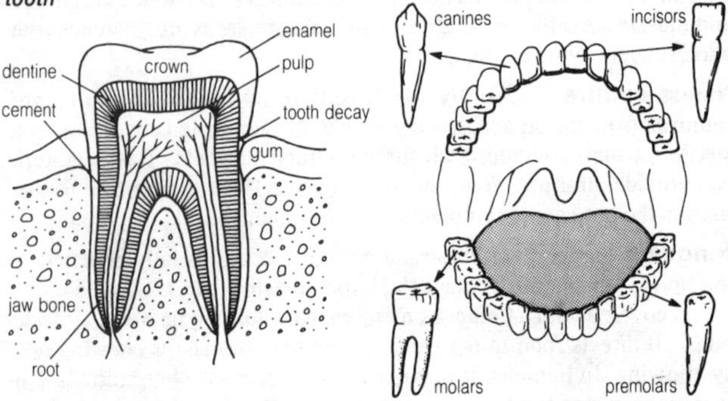

touch. Many animals, such as nocturnal ones, rely on touch more than humans do. Some have specialized organs of touch that project from the body, such as whiskers or antennae.

toxin or ***poison*** any chemical molecule that can damage the living body. In vertebrates, toxins are broken down by ◊enzyme action, mainly in the liver.

tracer a small quantity of a radioactive isotope used in ◊autoradiography to follow the path of a biological process. The location (and possibly concentration) of the tracer is usually detected by using a Geiger–Muller counter.

trachea in air-breathing vertebrates, the windpipe, the tube that conducts air from the larynx in the throat to the two bronchi (see ◊bronchus) that lead into the lungs. Like the bronchi, it is supported with rings of cartilage that prevent its collapse during breathing, and is lined with a ciliated mucous membrane that propels dust and other particles upwards towards the mouth.

In insects, the tracheae are large air-filled tubes that conduct air from the spiracles, small pores in the exoskeleton, into the body. Like vertebrate tracheae, they are supported by cartilaginous rings. The tubes extend throughout the insect, occasionally opening out into large air-filled sacs. The thinnest branches of the tubes are the ***tracheoles***, which have thin walls and are often filled with fluid. These smaller tubes are the true sites of gas exchange.

transcription in living cells, the process by which the information for the synthesis of a protein is transferred from the ◊DNA strand on which it is carried to the ◊RNA strand involved in the actual synthesis. It occurs by the formation of ◊base pairs when a single strand of unwound DNA serves as a template for assembling the complementary nucleotides that make up the new RNA strand.

transect in ecology, an imaginary line drawn across a piece of land, along which animal and plant diversity can be sampled. In fieldwork surveys the line can be made real by stretching a string between poles, and the diversity of species sampled by using a ◊quadrat. The data obtained may then be used to show how the wildlife varies in the area under study, and perhaps related to changes in abiotic and biotic conditions. For example, the progressive change in diversity (***zonation***) revealed by a transect survey of a beach, from the sand dunes to the high-tide mark, might be related to changing conditions of water exposure.

transfusion intravenous delivery of blood or blood products (plasma, red cells) into a patient's circulation to make up for deficiencies due to disease, injury, or surgery.

Blood transfusion was a highly risky procedure until the discovery of ◊blood groups indicated the need to check that the donated blood was compatible with that of the patient.

translation in living cells, the process by which proteins are synthesized. The information coded as a sequence of nucelotides in ◊RNA during ◊transcription is 'translated' in the ribosomes into a sequence of amino acids to form a peptide chain.

translocation the movement of soluble materials through ◊vascular plants.

Roots, stems and leaves all possess ◊vascular bundles, groups of hollow fibres that transport fluids and dissolved substances. Two types of tube exist within the bundles: ◊xylem for the upward transport of inorganic materials from root to leaf, and ◊phloem for the downward movement of organic substance formed during photosynthesis. Lower plants such as mosses lack these structures and are therefore less able to grow in dry areas.

transpiration the loss of water from a plant by evaporation. Most water is lost from the leaves through pores known as stomata (see ◊stoma), whose primary function is to allow ◊gas exchange between the internal plant tissues and the atmosphere. Transpiration from the leaf surfaces causes a continuous upward flow of water from the roots via the ◊xylem, which is known as the ***transpiration stream***. See also ◊xerophyte.

tree perennial plant with a woody stem, usually a single stem or 'trunk', made up of ◊wood, and protected by an outer layer of ◊bark. It absorbs water through a ◊root system. There is no clear dividing line between shrubs and trees, but sometimes a minimum height of 6 m is used to define a tree.

tricuspid valve a flap of tissue situated on the right side of the ◊heart between the atrium and the ventricle. It prevents blood flowing backwards when the ventricle contracts.

trophic level in ecology, the position occupied by a species (or group of species) in a ◊food chain. The main levels are ***primary producers*** (photosynthetic plants), ***primary consumers*** (herbivores), ***secondary consumers*** (carnivores), and ***decomposers*** (bacteria and fungi).

tropism or ***tropic movement*** the directional growth of a plant, or part of a plant, in response to an external stimulus. If the movement is directed towards the stimulus it is described as positive; if away from it, it is negative. ***Geotropism***, the response of plants to gravity, causes the root (positively geotropic) to grow downwards, and the stem (negatively geotropic) to grow upwards. ***Phototropism*** occurs in response to light, ***hydrotropism*** to water, and ***chemotropism*** to a chemical stimulus.

trypsin an enzyme in the gut responsible for the digestion of protein molecules. It is secreted by the pancreas but in an inactive form known as trypsinogen. Activation into working trypsin occurs only in the small intestine, owing to the action of another enzyme enterokinase, secreted by the wall of the duodenum. Unlike the digestive enzyme pepsin, found in the stomach, trypsin does not require an acid environment.

tuber swollen region of an underground stem or root, usually modified for storing food. The potato is a stem tuber, as shown by the presence of terminal and lateral buds, the 'eyes' of the potato. Root tubers lack these. Both types of tuber can give rise to new individuals and so provide a means of ◊vegetative reproduction.

Unlike a bulb, a tuber persists for one season only; new tubers developing on a plant in the following year are formed in different places. See also ◊rhizome.

tumour overproduction of cells in a specific area of the body, often leading to a swelling or lump. Tumours are classified as ***benign*** or ***malignant*** (◊cancers).

Benign tumours grow more slowly, do not invade surrounding tissues, do not spread to other parts of the body, and do not usually recur after removal. The most familiar types of benign tumour are warts on the skin.

turgor the rigid condition of a plant caused by the fluid contents of a plant cell exerting a mechanical pressure against the cell wall. Turgor supports plants that do not have woody stems.

turgor

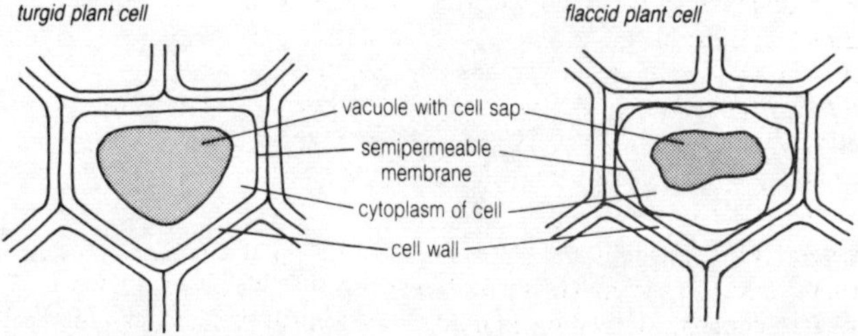

twin one of two young produced from a single pregnancy. Human twins may be genetically identical, having been formed from a single fertilized ovum (egg) that split into two cells, both of which became implanted. Nonidentical twins are formed when two ova are fertilized at the same time.

tympanic membrane or ***ear drum*** membrane capable of vibrating in response to sound waves passing from the outer ◊ear. The vibrations of the membrane are transferred to the tiny bones of the inner ear, which themselves pass vibrations through the inner ear and ◊cochlea.

U

ulna one of the two bones (the other is the radius) found in the lower limb of the tetrapod (four-limbed vertebrate).

umbilical cord in mammals, the connection between the ◊embryo and the ◊placenta. It has one vein and two arteries, transporting oxygen and nutrients to the developing young, and removing waste products. At birth, the connection between the young and the placenta is no longer necessary. The umbilical cord drops off or is severed, leaving a scar called the navel.

unicellular organism organism consisting of a single cell. Most are invisible without a microscope but a few, such as the giant ◊*Amoeba*, may be visible to the naked eye. The main groups of unicellular organisms are bacteria, protozoa, unicellular algae, and unicellular fungi, or yeasts.

urea ($CO(NH_2)_2$) waste product formed in the mammalian liver when nitrogen compounds are broken down. It is excreted in urine.

ureter tube connecting the kidney to the bladder. Its wall contains fibres of ◊involuntary muscle, whose contractions aid the movement of urine out of the kidney.

urethra in mammals, a tube connecting the bladder to the exterior. It carries urine, and in males, semen.

urinary system the system of organs that removes nitrogeneous waste products and excess water from the bodies of animals. In mammals, it consists of a pair of kidneys, which produce urine; ureters, which drain the kidneys; a bladder, which stores the urine before its discharge, and a urethra, through which the urine is expelled.

urine an amber-coloured fluid made by the kidneys from the blood. It contains excess water, salts, proteins, waste products in the form of urea, a pigment, and some acid.

uterus a hollow muscular organ of female mammals, located between the bladder and rectum, and connected to the Fallopian tubes above and the vagina below. The embryo develops within the uterus, and is attached to it, after implantation, via the ◊placenta and umbilical cord. The outer wall of the uterus is composed of involuntary muscle, capable of contracting powerfully when giving birth. Its lining changes during the ◊menstrual cycle.

V

vaccine any preparation of modified viruses or bacteria that is introduced into the body in order to induce the specific ◊antibody reaction that produces ◊immunity against a particular disease. It is may be introduced by mouth, by hypodermic syringe, or by means of a scratch on the skin surface.

vacuolation the development of a water vacuole inside plant cells. It brings about the enlargement of cells newly produced by cell division in the ◊meristem.

vacuole fluid-filled, membrane-bound cavity inside a cell. It may be a reservoir for fluids that the cell will secrete to the outside, or be filled with excretory products or essential nutrients that the cell needs to store. In the single-celled *Amoeba*, vacuoles are the sites of digestion of engulfed food particles. Plant cells usually have a large central vacuole for storage.

vagina the front passage in female mammals, linking the uterus to the exterior. It admits the penis during sexual intercourse, and is the birth canal down which the fetus passes during delivery.

valve a structure for controlling the direction of the blood flow. In humans and other vertebrates, the contractions of the beating heart cause the correct blood flow into the arteries because a series of valves prevent back flow.

variation the differences between individuals of the same species, found when examining a population. Such variations may be almost unnoticeable in some cases, obvious in others, and can concern many aspects of the organism. Typically, variation in size, behaviour, biochemistry, or colouring may be investigated. The cause of the variation may be genetic (and therefore inherited), environmental, or more usually a combination of the two. The origins of genetic variation can be traced to the recombination of the genetic material during the formation of the gametes, and, more rarely, to mutation.

variegation a description of plant leaves or stems that exhibit patches of different colours. The term is usually applied to plants that show white, cream, or yellow on their leaves, caused by areas of tissue that lack the green pigment ◊chlorophyll.

vascular bundle a strand of primary conducting tissue (a 'vein') in vascular plants, consisting mainly of water-conducting tissue, primary ◊xylem, and nutrient-conducting tissue, ◊phloem. It extends from the roots to the stems and leaves. Typically the phloem is situated nearest to the epidermis and the xylem towards the centre of the bundle. In plants exhibiting ◊secondary growth, the xylem and phloem are sepa-

vascular bundle

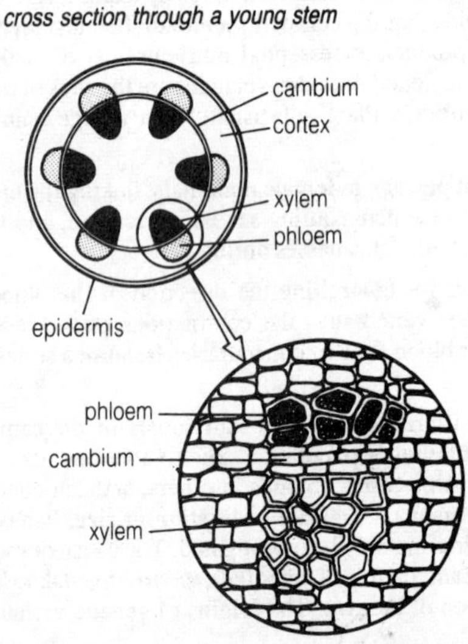

rated by a thin layer of vascular ◊cambium, which gives rise to new conducting tissues.

vascular plant a plant containing vascular bundles. Pteridophytes (ferns, horsetails, and club mosses), gymnosperms (conifers and cycads), and angiosperms (flowering plants) are all vascular plants.

vas deferens in male vertebrates, a tube conducting sperm from the testis to the urethra. The sperms are carried in a fluid (semen) secreted by various glands, and can be transported very rapidly when the involuntary muscle in the wall of the vas deferens undergoes rhythmic contractions, as in sexual intercourse.

vegetative reproduction a type of ◊asexual reproduction in plants that relies not on spores, but on multicellular structures formed by the parent plant. Some of the main types are ◊runners, sucker shoots produced from roots (see ◊suckering), ◊tubers, ◊bulbs, ◊corms, and ◊rhizomes. Vegetative reproduction has long been exploited in horticulture and agriculture, with various methods employed to multiply stocks of plants; see ◊propogation of plants.

vein in animals with a circulatory system, any vessel that carries blood from the body to the heart. Veins contain valves that prevent the blood from running back when moving against gravity. They always carry deoxygenated blood, with the exception of the veins leading from the lungs to the heart in birds and mammals, which carry newly oxygenated blood.

The term is also used more loosely for any system of channels that strengthens living tissues and supplies them with nutrients—for example, leaf veins (see ◊vascular bundle), and the veins in insects' wings.

vena cava one of the large, thin-walled veins found just above the ◊heart, formed from the junction of several smaller veins. The ***posterior vena cava*** receives oxygenated blood returning from the lungs, and empties it into the left atrium. The ***anterior vena cava*** collects deoxygenated blood returning from the rest of the body and passes it into the right side of the heart, from where it will be pumped into the lungs.

ventral in animals, term describing the lower surface, or the surface closest to the ground. The ventral surface of vertebrates is the surface furthest from the backbone; it faces forwards in bipedal (two-legged) vertebrates such as humans.

ventricle one of a pair of powerful pumping chambers in the lower half of the vertebrate heart. The ventricles are characterized by their thick muscular walls and for their dependence on the coronary artery. A heart attack (coronary thrombosis) occurs when the coronary artery is blocked by a clot, and the ventricles are denied an adequate supply of oxygenated blood.

venule small vein, found between the capillary beds and the larger veins. It contains deoxygenated blood at low pressure.

vertebra one of the small bones that make up the ◊vertebral column. They vary in form according to position and are capable of only limited movement.

vertebral column or ***spine*** the backbone, which gives support to an animal and protects the central nervous system. It is made up of a series of bones called ***vertebrae*** (26 in most mammals), running from the skull to the tail with a central canal containing the nerve fibres of the spinal cord. In tetrapods (four-legged vertebrates) the vertebrae show some specialization with the shape of the bones varying according to position. In the chest region the upper or thoracic vertebrae are shaped to form connections to the ribs. The backbone is only slightly flexible to give adequate rigidity to the animal structure.

In humans, there are seven cervical vertebrae, in the neck; 12 thoracic, in the upper trunk; five lumbar, in the lower back; the sacrum (consisting of five rudimentary vertebrae fused together, joined to the hipbones); and the coccyx (four vertebrae, fused into a tailbone).

vertebrate any animal with a backbone. The 41,000 species of vertebrates include mammals, birds, reptiles, amphibians, and fishes. They include most of the larger animals, but in terms of numbers of species are only a tiny proportion of the world's animals.

vestigial organ organ that remains in diminished form after it has ceased to have any significant function in the adult organism. In

vertebral column

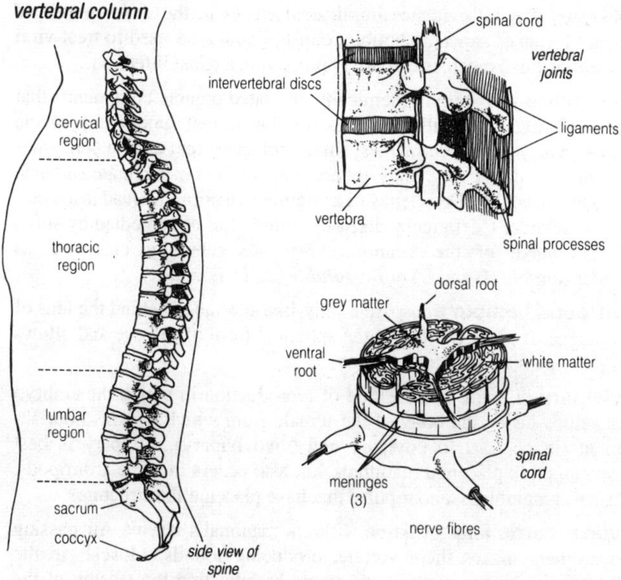

humans, the appendix is vestigial, having once had a digestive function in the ancestors of our species.

villus plural ***villi*** small finger-like projection extending into the interior of the small intestine and increasing the absorptive area of the gut wall. Digested foods, such as sugars and amino acids, pass into the villi and are carried away by the circulating blood.

virus an infectious particle consisting of a core of nucleic acid (DNA or RNA) enclosed in a protein shell. Viruses are acellular and are able to function and reproduce only if they can invade a living cell to use the cell's system to replicate themselves. In the process they may disrupt or alter the host cell. Among diseases caused by viruses are chickenpox, the common cold, herpes, influenza, rabies, AIDS, and many plant diseases. Recent evidence implicates viruses in the development of some forms of cancer. ◊Antibiotic drugs cannot be used to treat viral diseases, but ◊vaccines can offers protection against infection.

vitamin any of various chemically unrelated organic compounds that are necessary in small quantities for the normal functioning of the body. Many act as coenzymes, small molecules that enable ◊enzymes to function effectively. They are normally present in adequate amounts in a balanced diet. Deficiency of a vitamin will normally lead to a metabolic disorder ('deficiency disease'), which can be remedied by sufficient intake of the vitamin. They are generally classified as ***water-soluble*** (B and C) or ***fat-soluble*** (A, D, E, and K).

vitreous humour transparent jelly-like substance behind the lens of the ◊eye. It gives rigidity to the spherical form of the eye and allows light to pass through to the retina.

vivipary in animals, a method of reproduction in which the embryo develops inside the body of the female from which it gains nourishment (in contrast to ◊ovipary and ◊ovovivipary). Vivipary is best developed in placental mammals, but also occurs in some arthropods, fishes, amphibians, and reptiles that have placentalike structures.

vocal cords folds of tissue within a mammal's larynx. Air passing over them makes them vibrate, producing sounds. Muscles in the larynx change the pitch of the sound by adjusting the tension of the vocal cords.

voice sound produced by the passage of air between the ◊vocal cords. In humans the sound is much amplified by the hollow sinuses of the face, and is modified by the movements of the lips, tongue, and cheeks.

W

wall pressure in plants, the mechanical pressure exerted by the cell contents against tdhe cell wall. The rigidity (turgor) of a plant often depends on the level of wall pressure found in the cells of the stem. Wall pressure falls if the plant cell loses water.

water cycle in ecology, the natural circulation of water through the ◊biosphere. Water is lost from the Earth's surface to the atmosphere

water cycle

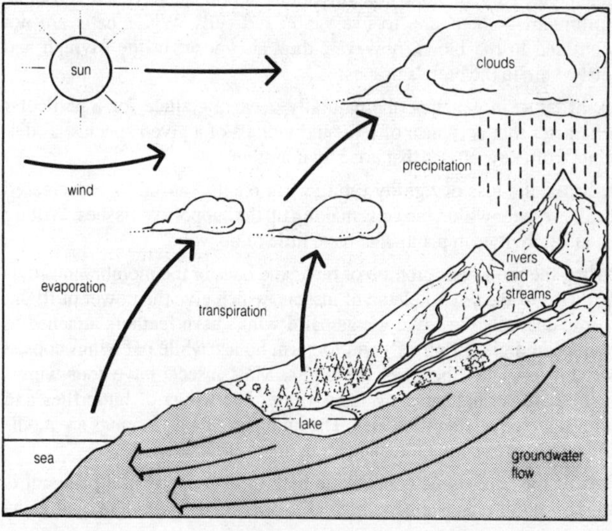

either by evaporation from the surface of lakes, rivers, and oceans or through the transpiration of plants. This atmospheric water forms clouds that condense to deposit moisture on the land and sea as rain or snow. The water that collects on land flows to the ocean in streams and rivers.

weedkiller another name for ◊herbicide, a chemical that kills some or all plants.

white blood cell or ***leucocyte*** one of a number of different cells that play a part in the body's defences and give immunity against disease. Some (◊phagocytes and ◊macrophages) engulf invading micro-organisms, others kill infected cells, while ◊lymphocytes produce more specific immune responses. White blood cells are colourless, with clear or granulated cytoplasm, and are capable of independent amoeboid movement. Unlike mammalian red blood cells, they possess a nucleus. Human blood contains about 11,000 leucocytes to the cubic millimetre—about one to every 500 red cells. White cells are not confined to the blood, however; they also occur in the ◊lymph and elsewhere in the body's tissues.

wild type in genetics, the naturally occurring allele for a particular character that is typical of most individuals of a given species, as distinct from new alleles that arise by mutation.

wilting the loss of rigidity (◊turgor) in plants, caused by a decreased wall pressure within the cells making up the supportive tissues. Wilting is most obvious in plants that have little or no wood.

wing the modified forelimb of birds and bats, or the membranous outgrowths of the ◊exoskeleton of insects, which give the power of flight. Birds and bats have two wings. Bird wings have feathers attached to the fused digits ('fingers') and forearm bones, while bat wings consist of skin stretched between the digits. Most insects have four wings, which are strengthened by wing veins. The wings of butterflies and moths are covered with scales. The hind pair of a fly's wings are modified to form two knoblike balancing organs (halteres).

wood the hard tissue beneath the bark of many perennial plants; it is composed of water-conducting cells, or secondary ◊xylem, and gains its hardness and strength from deposits of ◊lignin. ***Hardwoods***, such as oak, and ***softwoods***, such as pine, have commercial value as structural material and for furniture.

The central wood in a branch or stem is known as ***heartwood*** and is generally darker and harder than the outer wood; it consists only of dead cells. As well as providing structural support, it often contains gums, tannins, or pigments which may impart a characteristic colour and increased durability. The surrounding ***sapwood*** is the functional part of the xylem that conducts water.

The ***secondary xylem*** is laid down by the vascular ◊cambium which forms a new layer of wood annually, on the outside of the existing wood and is visible as an ◊annual ring when the tree is felled.

KH

X

X chromosome the larger of the two ◊sex chromosomes, the smaller being the ◊Y chromosome. In female mammals, the X chromosome is paired with another X chromosome; in males, it is paired with a Y chromosome. Genes carried on the X chromosome produce the phenomenon of ◊sex linkage.

xerophyte a plant adapted to live in dry conditions. Common adaptations to reduce the rate of ◊transpiration include a reduction of leaf size, sometimes to spines or scales; a dense covering of hairs over the leaf to trap a layer of moist air (as in edelweiss); and permanently rolled leaves or leaves that roll up in dry weather (as in marram grass). Many desert cacti are xerophytes.

xylem a tissue found in ◊vascular plants, whose main function is to transport water and dissolved mineral nutrients from the roots to other parts of the plant. In angiosperms, xylem is composed of a number of different types of cell, including the continuous conducting vessels, fibres (◊schlerenchyma), and thin-walled ◊parenchyma cells.

Non-woody plants contain only primary xylem, derived from the procambium, whereas in trees and shrubs this is replaced for the most part by secondary xylem, formed by ◊secondary growth from the actively dividing vascular ◊cambium. The cell walls of the secondary xylem are thickened by a deposit of ◊lignin, providing mechanical support to the plant; see ◊wood.

Y

Y chromosome the smaller of the two sex chromosomes. In male mammals it occurs paired with the other type of sex chromosome (X), which carries far more genes. The Y chromosome is the smallest of all the mammalian chromosomes and is considered to be largely inert (that is, without direct effect on the physical body).

yeast one of various sincle-celled fungi that form masses of minute circular or oval cells by budding. When placed in a sugar solution the cells multiply and convert the sugar into ethanol (alcohol) and carbon dioxide; see ◊anaerobic respiration. Yeasts are used as fermenting agents in baking, brewing, and the making of wine and spirits.

yolk a store of food, mostly in the form of fats and proteins, found in the ◊eggs of many animals. It provides nourishment for the growing embryo.

Z

zonation progressive change in the diversity of wildlife along a ◊transect (imaginary or real line) from one end of a habitat to another. On a rocky shore, for example, the organisms nearest to the low-tide mark, such as mussels, show adaptations that enable them to survive both immersion in the sea and exposure to the air; organisms on the the cliffs or sand dunes are adapted to survive sea spray, but cannot live for even a short time under water.

zone of elongation area within a root where the cells enlarge by ◊vacuolation. It is a region of rapid extension.

zoology the branch of biology concerned with the study of animals. It includes description of present-day animals, the study of evolution of animal forms, anatomy, physiology, embryology, behaviour, and geographical distribution.

zygote an ◊ovum (egg) after ◊fertilization but before it undergoes cleavage to begin embryonic development.

The Hutchinson Pocket Encyclopedia 1994 edition

The Hutchinson Pocket Encyclopedia packs as much information into its 640 pages as other volumes costing half as much again. Written in clear, jargon-free language, it covers all areas of knowledge from science and engineering to biographies of major sporting, cultural, and political figures, to newly updated information on all the countries of the world.

With more than 10,000 up-to-date entries, 150 maps and diagrams, and a host of useful chronologies and tables, the Pocket is quite simply the most comprehensive compact encyclopedia available.

ISBN 0 09 178286 4

Hutchinson Pocket Dictionary of Computing

If you use or are considering buying a personal computer, this dictionary is for you. In a compact format, here are all the terms that confuse, obscure, and mystify, made clear and intelligible to the non-expert. Included are all the latest terms – image compression, local bus, MPC – as well as basic topics such as boot disc and graphics card. Whether you are a newcomer to computing or a seasoned hand, you will find this dictionary an essential aid.

published October 1993

ISBN 0 09 178104 3

Hutchinson Pocket Chronology of World Events

From early humans in Asia (800,000 BC) and the construction of the Pyramids (2500 BC) to the breakup of the USSR (Dec 1991) and the 1992 Earth Summit, the Pocket Chronology lists the major events in world history, including science, politics, religion, and society. Here is the whole of world history at a glance.

published October 1993

ISBN 0 09 178275 9

[redacted] PUB - 6TH FORM.

PHIL BATES 13B.

Hutchinson Pocket Dictionary of Quotations

From Aristotle to Yeltsin, this dictionary of more than 3,000 quotations is fully indexed to enable quotations to be found from individual words within the quotation. Historical, literary, political, topical, and humorous quotations are all included, resulting in a book that can be browsed through for hours.

published October 1993

ISBN 0 09 178281 3